Climate Change and Writing the Canadian Arctic

Renée Hulan

Climate Change and Writing the Canadian Arctic

palgrave
macmillan

Renée Hulan
Saint Mary's University
Halifax, Nova Scotia, Canada

ISBN 978-3-319-69328-6   ISBN 978-3-319-69329-3   (eBook)
https://doi.org/10.1007/978-3-319-69329-3

Library of Congress Control Number: 2017961765

Cover illustration: Pattern adapted from an Indian cotton print produced in the 19th century

Printed on acid-free paper

This Palgrave Pivot imprint is published by Springer Nature
The registered company is Springer International Publishing AG
The registered company address is: Gewerbestrasse 11, 6330 Cham, Switzerland

"Lear of Whylah Falls" by George Elliott Clarke, from the publication *Whylah Falls* published by Gaspereau Press. Used by permission of the publisher.

Poems in *Terror and Erebus* from *Afterworlds* by Gwendolyn MacEwen published by McClelland. Used by permission of the estate of Gwendolyn MacEwen.

Preface: Writing the Anthropocene

We live in what Amitav Ghosh calls *The Great Derangement*, a time when it has become clear that "global warming is in every sense a collective predicament," yet "humanity finds itself in the thrall of a dominant culture in which the idea of the collective has been exiled from politics, economics, and literature alike" (80). From the plight of the Sundarbans that Ghosh describes to Pangnirtung, Nunavut, and Kivalina, Alaska, the effects of global warming are devastating communities, causing mass destruction, and displacing populations. In the film *Qapirangajuq: Inuit Knowledge and Climate Change*, elders in Pangnirtung recount the 2008 flood that occurred when the river changed course, a once in a lifetime occurrence that the people attribute to climate change. The story of Kivalina, an Inupiaq village that is rapidly falling into the Chukchi Sea is recounted in Christine Shearer's *Kivalina: A Climate Change Story*. As early as 1990, members of the community were asking to be relocated as they watched an island that had covered approximately 55 acres in 1953 shrinking: by 2003, it covered only 27 acres. According to Kivalina Tribal Administrator Colleen Swan, the Inupiaq had been noticing changes in the environment for decades but their genius for adapting had been frustrated by forced settlement and the end of their nomadic way of life (Shearer 76). Once protected by the sea ice, the village is now vulnerable to erosion as the ice forms later and melts earlier each year. In 2006, the seawall protecting the village was washed away, and in 2007, another storm forced the evacuation of the whole community (Shearer 15).

Even in the face of such drastic events, climate change skeptics continue to insist that global warming is not real. To understand how they succeed in convincing decision makers, Shearer compares the tactics of these skeptics to those used by the asbestos, lead, and tobacco industries. In each case, science has been enlisted to support the rejection of scientific data. Each of these industries has funded their own private studies of the harmful effects of their products while working to suppress, dispute, and deny independent research, but their most successful tactic is using scientific method against itself. The very process of "gathering and assessing reliable data, producing replicable results, establishing areas of consensus, and building on the findings for greater understanding, while acknowledging areas in need of future research" could be presented by skeptics as "uncertainty" and as an argument for actually rejecting the scientific conclusions reached (Shearer 16). Similarly, climate change deniers cite "certainty as the only acceptable standard for acknowledging and thus acting on climate change while simultaneously manufacturing uncertainty to make sure certainty is never achieved" (Shearer 16). Meanwhile, Kivalina is falling into the sea.

An accomplished writer and observer of the novel as a genre, Ghosh surveys the contemporary literary scene and wonders why the subject of climate change has been exiled from "the mansion in which serious fiction has long been in residence" to the "generic outhouses" (24) occupied by science fiction, the Gothic, or the more recently designated "cli-fi." As a genre, cli-fi descends from a tradition born of a climate event when Mary Wollstonecraft Shelley, holed up in a villa beside Lake Geneva during the cold, wet "Year without a Summer" created *Frankenstein*, but the themes her novel explores were quickly separated from the literary mainstream as scientific and literary knowledges became estranged by the 'partitioning' of Nature and Culture within Modernity (Latour qtd. in Ghosh 66–68). As this estrangement grew, the bourgeois novel was beginning to displace poetry as the dominant form of literature, and realism, with its premise that the truth of small things should underwrite larger truths, emerged as its dominant mode, banishing what could not be readily verified by the reader's experience. Towns falling into the sea; ground disappearing beneath one's feet—such fantastic events could not be the stuff of realist fiction. As Ghosh observes, "the novel was midwifed into existence around the world, through the banishing of the improbable and the insertion of the everyday" (17). As the everyday came to replace the incredible in storytelling, and probability came to authorize representation of all kinds, it produced "the irony of the 'realist novel': the very gestures with which it conjures up reality are actually a concealment of the real" (23). The realist mode of the bourgeois novel, with its focus on the "moral adventure" of the individual continues to reproduce itself, dominating literary fiction worldwide, just as global capitalism and globalization dominate other spheres of life. Looking ahead to the future, Ghosh predicts that future generations will look back and conclude that "ours was a time when most forms of art and literature were drawn into the modes of concealment that prevented people from recognizing the realities of their plight" (11).

The two writers who are the focus of this book share concern for this plight yet come from different backgrounds. As an Inuit woman, Sheila Watt-Cloutier brings her knowledge of Inuit traditions and lived experience into the international world of diplomacy and public policy. As she takes the world stage to advocate for action on climate change, she maintains a worldview rooted in Inuit culture and Indigenous knowledge. In *The Right to Be Cold*, she recalls the struggle to move climate change to the forefront of the international agenda, and in the tradition of Inuit autobiography, she tells the story of an individual in relation to community. *The Right to Be Cold* argues that the protection of the ways of life depending on the Arctic environment, particularly the cold, is a human right that transcends national boundaries. Watt-Cloutier documents the role Indigenous peoples are playing in bringing attention to climate change and negotiating measures to deal with it at the international level. In these negotiations, Indigenous leaders from the Canadian Arctic are exerting their own sovereignty and forging alliances with others from the circumpolar world as well as with Small Islanders from around the globe to pressure governments to take action to protect Indigenous communities and traditions threatened by climate change.

The emerging circumpolar view of the Arctic in Canadian writing is also explored in Kathleen Winter's *Boundless*, her memoir of cruising the Northwest Passage. As the title signals, Winter is fascinated by "the possibility of 'belonging' outside borders" (Winter 56); so what begins as an adventure, an expedition through the fabled Northwest Passage on a Russian icebreaker becomes an attempt to understand what she calls "true sovereignty," the authenticity residing in the land itself as she experiences an initiation that involves unlearning and learning. Winter begins with a personal quest to cure the malaise of adulthood and to rediscover the imminence in the ordinary world she once glimpsed as a child, but what she gains is more than an appreciation for the natural world: it is an understanding of the relationship between the future of the Arctic environment and the future of Indigenous peoples. Winter's Arctic excursion causes her to reflect on her own place as a settler on Indigenous land and inspires her to learn more about the struggle of Indigenous people to protect their land. As a tourist, she participates in the commodification of the Arctic, yet resists the reproduction of adventure narratives even as she is caught up in one of her own. At the end of the book, she is still learning and still searching, even as she advocates for an Arctic under threat.

In these two examples, I see the beginning of a discourse based on knowledge. I also see the memoir as a form that, unlike the novel, *can* write the Anthropocene. Although the memoir can be used to craft and brand an image or to gatekeep around an author's sense of authority, the form can also voice the collective concerns of an author's community. Most importantly, as a form of non-fiction, the memoir can include what the novel, and other fictional forms, must conceal: data, facts, evidence. Published in 2014 and 2015 respectively, these memoirs exemplify the changing ways of imagining the Arctic that are being shaped by the modern and traditional knowledge of the Indigenous people who call it home.

My thinking about writing about the Canadian Arctic from a circumpolar perspective has benefited greatly from the support of the Arctic Modernities Research project funded by the Norwegian Research council and led by Anka Ryall at the University of Tromsø. For many enriching conversations over the years, I wish to thank Anka and the members of the group: Heidi Hansson, Fredrik Chr. Brøgger, Hanna Eglinger, Susi K. Frank, Sigfrid Kjeldaas, Johan Schimanski, Roswitha Skare, Peter Stadius, Ulrike Spring, Kirsten Thirsted, and Henning Howlid Waerp. And to Michael and Rebecca, all my love.

CONTENTS

CHAPTER 1

Franklin's Long Shadow: Representations of the Canadian Arctic

Abstract In an extensive analysis of visual imagery depicting the search for the HMS *Erebus* in 2014 and contrasting it with the search for the HMS *Terror* in 2016, Hulan introduces the idea that climate change is gradually transforming how Canadians write the Arctic. While the North is still viewed as a source of national and territorial sovereignty in Canada, climate change is creating a shift from nationalist to global perspectives on the Arctic. The importance of Indigenous knowledge is highlighted in the literature by tracing changing attitudes toward Inuit testimony concerning the expedition.

Keywords Northwest Passage • Franklin • Inuit testimony • HMS *Terror* • HMS *Erebus*

When the discovery of the HMS *Terror*, one of the two ships lost in the Arctic under Sir John Franklin's command, was announced on September 12, 2016, the British newspaper *The Guardian* was the first to run the story with a link to the video tour made by Adrian Schimnowski, the operations manager from the Arctic Research Foundation (ARF), a charitable organization founded by Canadian entrepreneur Jim Balsillie (www.theguardian.com/world/2016/sept/12/hms-terror-wreck-found-artctic-nearly-170-northwest-passage-attempt). The article quoted Schimnowski's description of the

© The Author(s) 2018
R. Hulan, *Climate Change and Writing the Canadian Arctic*,
https://doi.org/10.1007/978-3-319-69329-3_1

crew's first glimpses of the ship's interior: "We have successfully entered the mess hall, worked our way into a few cabins and found the food storage room with plates and one can on the shelves ... We spotted two wine bottles, tables and empty shelving. Found a desk with open drawers with something in the back corner of the drawer." These traces of human habitation, relics of exploration awaiting the tourist gaze, marked the underwater resting place of the HMS *Terror* and reunited it with its sister ship HMS *Erebus* which had been found in 2014.

As visual representation, the ARF video could not have differed more from "Great Canadian North," a glossy Government of Canada (GOC) advertisement that was broadcast in the months leading up to the 2015 federal election. The GOC led by the Conservative party commissioned the minute-long video inviting Canadians to join in the commemoration of the 150th anniversary of Confederation. The video begins with an aerial shot of ice floes on a vast and otherwise empty ocean. Faint throat singing can be heard in the background. There is a close-up of an Inuit man looking up at the sky, while another off in the distance shades his eyes to scan the horizon. In the next shot, two Inuit clad in sealskin clothing walk along carrying kayaks on their shoulders as the voiceover begins: "170 years ago, the inhabitants of the Arctic encountered explorers from another world embarked on a quest to find the Northwest Passage." The scene changes to a view of two nineteenth-century vessels, perhaps the *Terror* and the *Erebus*, sailing in water scattered with ice while instrumental music plays in the background. We see a sailor ringing the ship's bell as the captain looks through a brass telescope out to sea, and the voiceover continues: "Sir John Franklin's expedition was lost. But his disappearance launched an era of exploration unparalleled in Arctic history." Names like "Baffin" gradually appear on a sepia-colored map. In the next scene, a bearded white man wearing sealskin clothing is seen speaking to a young Inuit boy, an Inukshuk in the distance, as they are joined by an Inuit girl who looks out and points the way. As images of a bush plane, a snowmobile, a research station, a satellite, a helicopter, and a Canadian Coast Guard icebreaker follow, the voiceover concludes: "Franklin's legacy is one of perseverance, discovery, and innovation that lives on today and has helped to keep our True North strong, proud and free." The last image shows two divers underwater, the Parks Canada logo clearly legible on their suits as they shine a flashlight on the ship's bell and the narrator invites viewers to celebrate the anniversary of Confederation and "our great legacy of discovery" (www.Canada.pch.gc.ca/eng/14353195587305).

The history of the Arctic presented in this video follows a linear progression from ancient times to the present mapped by changing technologies of travel and communication as each scene offers a more recent mode of exploration. The kayak is superseded by sailing ships: the bush plane soaring overhead leads to the satellite circling the earth, the snowmobile and the icebreaker follow. Although machines are the protagonists, the human figures play the stereotyped role of Explorer. Inside the bush plane, a grizzled pilot surveys below; another male pilot looks out from the helicopter, his face mostly obscured by glasses, helmet, and microphone; the Coast Guard captain, like Franklin and the anonymous Explorer, is bearded and in uniform. Inuit characters occupy the moment of contact, dated arbitrarily at "170 years ago," and return to guide another bearded, white, male Explorer figure before vanishing from the narrative. John Franklin and the anonymous sailor are mirrored in the members of the Coast Guard and Parks Canada who carry on the work of discovery.

The symmetry of these scenes is created by images of vision: first an Inuit man looks skyward as another scans the horizon; Franklin looks through his nineteenth-century glass; the member of the Coast Guard lifts a pair of binoculars, and the two underwater divers use a flashlight the better to see the newly found Franklin relics beneath the sea. By representing discovery and exploration through images of looking, seeing, and searching, and those technologies that improve and enhance vision, the video emphasizes it above other senses even though, perhaps ironically, the discovery of the ship it celebrates was achieved with technologies that used sound to explore the seabed. The emphasis on vision, however, makes the short video the kind of "Arctic spectacle" that Russell Potter has shown was a popular theme in the visual art and panoramas of the nineteenth century. The Franklin expedition itself was equipped with a camera, the very latest in technology, and the searches for Franklin were depicted in numerous visual forms. In literal and figurative terms, vision characterizes the role the Arctic played in the Victorian imagination and now plays in Canadian culture for "it was principally through the technologies of vision that the Arctic was most keenly and energetically sought" (Potter 2007, 4).

If the identification of "perseverance, discovery, and innovation" with the Franklin expedition seems an unlikely one, it was not an unusual move for the Canadian government adept at bending historical narrative to its own purpose. The Franklin expedition was last seen in 1845, 22 years before Confederation, making its connection to the 2017 anniversary tenuous at best. In symbolic terms, the Franklin expedition, long regarded as the epitome of failure in technological and scientific progress—the limit reached by Modernity—seems an unlikely story for national celebration in Canada whereas in Britain, it served a particular national and imperial narrative of sacrifice. The Government of Canada attempted to graft the British legacy of heroic male sacrifice unto Canadian history, to write over the Canadian literary history of demythologizing exploration and discovery, and to recast Indigenous people as accomplices in the heroic narrative of Arctic exploration. As it had during the War of 1812 celebrations, the Conservative-led government's historical revisionism combined an excess of memory with strategic forgetting. It was a strategy that had served to identify the government with patriotism and heritage in those circles keen to see themselves as inheritors of British institutions and traditions (see Hulan 2015). By cherry-picking useful details, the public commemoration could be used to reflect the sort of Canada that the Conservative-led government wanted to promote as it campaigned for reelection. After the 2015 election, however, the video was taken down from the internet (www.Canada.pch.gc.ca/eng/14353195587305).

THE EXPLORER AND THE DOG CHILDREN

When they first encountered European explorers, some say the Inuit were astonished by the hairiness of their faces, and the Inuktitut word, Qalunaat, has been variously translated as referring to this response. Sheila Watt-Cloutier defines it as "the Inuktitut plural term for white people. It derived from *qallunaq*, which describes the bones on which the eyebrows sit, which protrude more on white people than on Inuit" (Watt-Cloutier 4). The glossary to *Sanaaq: An Inuit Novel* by Mitiarjuk Nappaaluk indicates: "White man, literally 'big eyebrows'" (Nappaaluk 216). The epigraph to Mini Aodla Freeman's *Life Among the Qalunaat* defines "Qallunaaq (singular)" and "qallunaat (plural)" as "literally 'people who pamper their eyebrows'; possibly an abbreviation of qallunaaraaluit: powerful, avaricious, of materialistic habit, people who tamper with nature" (Freeman np). In *Qalunaat! Why White People are Funny*, Zebedee Nungak describes 'qalunaat' as more of an attitude or worldview. What may at one time have been the description of a characteristic of some non-Indigenous people, maybe even an insider's joke at their expense, is now understood as the relative and relational term Emilie Cameron uses in *Far Off Metal River*

(13–17). As Cameron explains, the Inuktitut word "qablunaq" was never associated with skin color and therefore is inaccurately translated as "white person" though it is often used as such, as in Nungak's film; instead, it emerges from and encompasses Inuit experiences of encounter with non-Indigenous people through history from early traders and explorers to corporate and government officials. As such, it conveys the power relations defining colonialism.

While it is certain that the meaning of Qalunaat is not static and that it is only made meaningful in context, there is an intriguing connection between the legendary hairy-faced Europeans and the story of Nuliajuk. Also known as Sedna, Tullayoo, Uinigumasuittuq, she is the Mother of the Sea Beasts and of human beings.[1] The story continues to be told throughout the circumpolar world by Inuit including the young artist Ruben Komangapik who incorporates a QR code in his beautiful sculpture *Tigumiatuq* held by the National Gallery of Canada. The code takes the viewer to a video in which he tells the story of Nuliajuk (https://www.youtube.com/watch?v=dL2g8Sj0jRQ). The story has also been recorded many times by *qalunaat* visitors, and these versions have circulated throughout the world in written form. Komangapik's online telling is very similar to the one recorded by Franz Boas in *The Central Eskimo* on Baffin Island:

> Once upon a time there lived on a solitary shore an Inung [Inuk] and his daughter Sedna. His wife had been dead for some time and the two led a quiet life. Sedna grew up to be a handsome girl and the youths came from all around to sue for her hand, but none of them could touch her proud heart. (qtd, in Petrone 42)

This Sedna is lured into marriage by the deceitful fulmar. When her father comes to rescue her and to take her home, he kills the fulmar-husband, angering the rest of the flock who whip up the seas around the boat:

> In this mortal peril the father determined to offer Sedna to the birds and flung her overboard. She clung to the edge of the boat with a death grip. The cruel father then took a knife and cut off the first joints of her fingers. Falling into the sea they were transformed into whales, the nails turning into the whalebone. Sedna holding on to the boat more tightly, the second finger joints fell under the sharp knife and swam away as seals ... when the father cut off the stumps of the fingers they became ground seals. (qtd. in Petrone 42)

In some versions, Sedna then sinks to the bottom of the sea where she presides over the sea beasts; in this version, she is lifted into the boat and later revenges herself against her father by having her dogs attack him as he sleeps. In others, Sedna-Nuliajuk refuses to marry, so her father marries her to his dog. Knud Rasmussen also records several references to Sedna in volumes of *The Fifth Thule Expedition* later popularized in his book *Across Arctic America*. Traveling the Arctic in the early twentieth century, he documented the fear her name inspired. In his film *Nuliajuk: Mother of the Sea Beasts*, Canadian art dealer John Houston narrates a scene in which Rasmussen describes a shaman shaking as he sketched a drawing of Sedna. Like his hero Rasmussen, Houston is fascinated by the lasting power of Nuliajuk who he celebrates as "one of the last survivors from that time when all our ancestors worshiped mighty female gods." The film highlights Nuliajuk's role as the mother of creation. In some of the stories told in the film, Nuliajuk is fooled by a dog disguised as a beautiful young man. When her children are born, she fashions boats out of gigantic sealskin boots, places her children in them, and sets them adrift: those resembling human beings become the Indigenous people of the Americas, those resembling dogs become white people.

Imagining himself as one of the dog children returning home to the Arctic, his bearded face peering out from behind an actor dressed as the fulmar, Houston looks for the traces of the story and finds them everywhere, with each retelling, he sees the Inuit oral tradition surviving and transforming as a living culture. Dorothy Harley Eber recounts a version heard in the 1990s in which there are six children, two of them Inuit and the others half-dog and half-Indian; or, half-white and half-dog (Eber 21). In the GOC video, all of the individual figures are male, except for one young Inuit girl who points the way for a European explorer, and bearded like the veritable Dog children of Inuit oral tradition. This retelling of the Franklin disaster captured nothing of the continuity of Inuit culture in the Arctic revealed in the story of the Mother of the Sea Beasts, but instead casts Inuit as inhabitants of pre-contact history. The characters seen searching and "discovering" are uniformly male, and by putting beards on these characters, the advertisement further differentiated the explorers from the Inuit, thus visually attributing "perseverance, discovery, and innovation" exclusively to white men.

In 2014, Inuit stories were absent from the narrative as a clean-shaven Stephen Harper, the then Prime Minister of Canada, stepped into the Explorer role and appeared on television proclaiming triumphantly: "We found the ship!" The expedition in search of the remains of the HMS *Erebus* had been a favorite project for a Prime Minister often described as a "history buff" and an enthusiast of the North. The search was led by the highly respected archeologist Robert Grenier and made use of the Coast Guard icebreakers *Amundsen* and *Sir Wilfrid Laurier*. In announcing a federal grant of $75,000 to the project, the then Minister of the Environment, John Baird had exclaimed, "I am sure every historian, archae-ologist, and storyteller is as excited about this as I am," and added that the story had "the allure of an Indiana Jones mystery" (*Canada Launches New Arctic Search*). If Baird's allusion to the Spielberg film was meant to jazz up his presentation, it also betrayed the true audience for the press conference and pointed to the way in which nostalgia for British imperialism provides cover for serving multinational interests, including those of American cor-porations. At the same press conference, Baird also claimed that the Franklin expedition would confirm Canada's "long-standing presence in the Arctic" and thus "enhance issues of sovereignty," as if heeding advice from Franklyn Griffiths that "a mobilizing vision or guiding sense of national purpose" would be needed for Canada to gain and maintain control over "a safe and open Northwest Passage" (Griffiths 21).

The history of the North as a source of *Sovereignty or Security* for Canada is fully explored in Shelagh Grant's study by that title as well as in the work of Michael Byers, Rob Huebert, and P. Whitney Lackenbauer (Griffiths et al. 2011). But, as Grant's "Myths of the North in the Canadian Ethos" demonstrates, the representation of the North as a vital national heritage is only one lens through which the Canadian North has been viewed. The idea that the North gives "a distinct identity to the Canadian nation, its people, and its institutions" (Grant 1989, 39) has always been contested. While its construction, that is, the idea of "Canada as North," has been thoroughly dissected by historians, literary scholars, and cultural critics, myself included (see also Grace 2001; Sangster 2016), the myth of the North as a frontier to immeasurable natural wealth continues to animate arguments for northern heritage, a vision that Thomas Berger's report on the MacKenzie Valley Pipeline *Northern Frontier, Northern Homeland* was first to criticize by acknowledging the different perceptions of insiders and outsiders, northern inhabitants and southern interests. Indeed, when the *HMS Erebus* was found, it was heralded by the GOC as proof of Canada's Arctic sovereignty, prompting Adriana Craciun to write in the *Ottawa Citizen*: "Franklin's legacy is not discovery or sovereignty, it is disaster. That is the sobering lesson we should take from the heroic age of explora-tion" (Craciun 2016, 225).

The expedition that located the HMS *Erebus* not only claimed the expedition as part of "Canadian" history but recycled a nationalist rhetoric of northern sovereignty based on northern heritage and development that dedicated experts in Northern Studies had long worked to discredit. In 1977, Berger's *Northern Frontier, Northern Homeland* introduced Canadians to the idea that the "true north" they sought to exploit remains the homeland of modern Indigenous people. That same year, Bruce Hodgins used the term "colonial" to describe Canada's relationship with the North (10), and Kenneth Coates' *Canada's Colonies* offered this precise description: "The extensive powers of the national bureaucracy, the continued reliance on federal subsidies, and frequent federal intervention in regional affairs all make plain the north's colonial status" (9). A few years later, Coates and Judith Powell dedicated *The Modern North: People, Politics, and the Rejection of Colonialism* (1989) to the decolonization of the Canadian North (xvi). Studies of the northern environment have also identified the destruction caused by rapid development of northern resources and analyzed social and political relations in the North.

Since this groundbreaking work, an awareness of the unequal power relations between the North and the rest of Canada has transformed both the research agenda and methodologies used to advance it in Northern Studies. In literary studies, the work of decolonization began with the debate about the ethics of speaking for others. Amid calls by Indigenous writers for Canadians to stop stealing Native stories, literary scholars had to choose between ignoring the literary production of Indigenous people and trying to find ethical ways of studying and teaching it. As Margery Fee explains, answers to these problems could be learned from Indigenous ways of knowing: "Treating others with respect, even those others we disagree with, by paying close attention to what they say, is an ethic worked out over generations by Indigenous people" (16). The practices of active listening and reading from the inside out guided literary scholars who want to support the work of decolonization. In Northern studies, Dale Blake and Robin McGrath took up this work by making Inuit literature in English available to readers while arguing vigorously for its inherent literary characteristics. More recently, Keavy Martin's *Stories in a New Skin* explores the role stories play in all aspects of life, including social organization, as "an articulation of peoplehood—of a uniquely Inuit humanity" (Martin 15). Learning from these traditions requires constant awareness of the difference between dialogue and cultural appropriation that

Martin clarifies by stressing the cultural value placed on reciprocity in Inuit culture. At the same time, a number of studies, such as Marybelle Mitchell's *From Talking Chiefs to a Native Corporate Elite: The Birth of Class and Nationalism among the Canadian Inuit* and Renée Fossett's *In Order to Live Untroubled: Inuit of the Central Arctic, 1550 to 1940*, have added to understanding of northern cultures, history, and politics. The creation of Nunavut in 1999 marked a movement toward greater autonomy and self-government and corresponded to a greater sense of sovereignty in social and cultural terms.

FINDING FRANKLIN, AGAIN AND AGAIN

In contrast to the territorial concept of sovereignty used by the Canadian state, Indigenous people worldwide use "sovereignty" to mean the recognition and assertion of their right to exist and to determine their own futures. In 2010, Parks Canada gestured toward Inuit sovereignty in promotional material related to the recent expeditions in search of Franklin's ships by stressing the involvement of the Government of Nunavut through its Department of Culture, Languages, Elders and Youth: "The expedition team has worked closely with the Inuit people in conducting research and in planning this search. It is widely accepted that Inuit oral history and research could hold the key to unlock the discovery of *HMS Erebus* and *Terror*" (www. pc.gc.ca). Members of the team cited the expedition's verification of oral testimony as found in David Woodman's *Unravelling the Franklin Mystery*, and Robert Grenier himself said that he hoped to make up for "the way, in his opinion, Franklin snubbed the Inuit" (qtd. in Edward 33). Nevertheless, what the discovery of the ships could offer to the citizens of Nunavut was never explicitly stated, though the possibility of training and employment in the archeological work supporting the undersea operation was cited as a benefit. Arctic history, however, suggests that any expedition will have unforeseen consequences. As Graeme Wynn shows, study of the materials discarded in search of Franklin's ships, such as those left behind by the *HMS Investigator* in the 1850s "reveals the ramifying social, economic, and often difficult-to-discern ecological consequences of even the most limited contact between Europeans and Indigenous people" (45). Not surprisingly, northern citizens are wary of southern interest driven as much by corporate interests as national ones, especially in anticipation of the impact of climate change. Announcements of increased military funding in 2004 and 2006 were received with caution and new calls for talks with Ottawa by the government of Nunavut. A *Nunatsiaq News* report headlined "Armed Forces to Pour More Money into North," the commander of the Canadian Forces Northern Area, Col. Normand Couturier, noted that, "[t]he North is very important to Canadian defense" and speculated that it would be "given priority because of the sovereignty issue" (Younger-Lewis 2004). But the coverage did not shy away from the underlying reason for southern concern to protect oil and diamonds that would become easier to access as the Northwest Passage remains free of ice for longer periods. As the new rush to stake out claims in the Arctic continues, Canada is taking steps to exert its sovereignty, especially in the Northwest Passage, which Canada claims is an internal waterway. Thus, at the beginning of the new millennium, there was a brief return of northern nationalism, especially regarding the Arctic, with rhetoric steeped in the language of the heroic age of discovery and exploration in the search for the Northwest Passage. At the same time, the nationalism inflecting representations of the North, including the Arctic, seems out of place in a global crisis, and the addition of the concept of stewardship to more recent discussions of sovereignty and security in Canada seems to signal recognition of the environmental dangers inherent in unfettered development. Two competing discourses have emerged: one that seeks to protect Arctic sovereignty for economic development, and the other that seeks to protect the sustainable ways of life in the Arctic against environmental devastation. Both discourses are responding to rapid change in the Arctic.

The search for Franklin's ships tells a story of how colonial relations in the North function as the national interest in Canadian sovereignty and security continues to shape local and regional life powered by the GOC's ability to fund research and to enter into collaboration with corporations like Shell. Indeed, there is a long history of "Arctic opportunism" (Craciun 2016, 229) in Canada that is seeing its most recent manifestation in the treatment of the HMS *Terror* and the HMS *Erebus* as artifacts that cement Canada's claims. Craciun links the former GOC's handling of the announcement in 2014 to the "elimination of archaeologists working throughout the Canadian Arctic, the closure and consolidation of regional archives, and the shutdown of research revealing the multicultural richness of the Arctic five hundred years before the English arrived," not to mention the censorship of scientists employed by the GOC that betrayed its hostility toward both history and science (Craciun 2016, 229). "In this version of Arctic opportunism, the Arctic is emptied of Indigenous people and its multiplicity of histories, and reduced to an environment full of energy awaiting extraction" (Craciun 2016, 230).

At the same time, Craciun notes that "Canada's territorial sovereignty is not in dispute" (Craciun 2016, 225), unlike its claim that the Northwest Passage is an internal waterway. What may have seemed like, and surely was, nationalist propaganda to observers also spoke to the potential threat to Canada posed by an increased international presence in the Arctic. This specter had been raised by the Minister of the Environment Baird in relation to foreign access to natural resources, especially minerals. It may have been a phantom fear; however, it was accompanied by real political action, including spending on an increased military presence patrolling the region. Prime Minister Harper was photographed after a ride in one of Canada's F-18s, conjuring images of Russian and American fighter jets to remind Canadians of their position relative to their neighbors' military might. While the Conservative-led government had modified the colonial approach taken in the past by including local communities, those communities, in this case in Nunavut, remained in the position of colonies accepting decisions made elsewhere and living with a set of priorities that reflects national, and sometimes private, interests.

It may be too soon to know if the announcement on September 12, 2016 of the discovery of the HMS *Terror* in Terror Bay off King William Island signals a change in approach to the Arctic by the GOC; however, in highlighting the role of Inuit participants, it showed a marked difference from the handling of the 2014 discovery of the *Erebus*. With its lack of polish and performance, the ARF video of the 2016 discovery of the HMS *Terror* gives the impression of an excited amateur attempting to capture the moment: the natural light on the overcast day produces a murky effect, and the camera does nothing to disguise the motion of the research vessel *Martin Bergmann* beneath it. The video was shot and produced by Schimnowski who introduces "the first images that we captured on September 3 at 8:30 in the morning." With Stan Roger's "Northwest Passage" playing softly in the background, the crew members introduce themselves as the camera pans the underwater scene. The overall effect evokes an uncut, unedited piece of footage gathered to document scientific work. Each speaker's brief statement roughly matches the expertise they bring to the expedition: the captain of the vessel identifies the double-wheeled helm; a female graduate student researching the physics of the Arctic ocean describes the bow's wooden planks on the outside of the vessel and the steel sheeting on the ice guard. One by one, the crew members explain what the viewer sees: the capstan and rope on the port side; the ship's bell described as the "heart and soul of the HMS *Terror*"; the hatch leading to the mess hall and table inside; the exhaust pipe for the steam engine; there is a shot of the cabin with the crew showing the drawings used to confirm the find. About halfway through the video, the tone changes as Sammy Kogvik from Gjoa Haven tells his story: "Six, seven, eight years ago, I was on my way to the lake to go put our nets out when we got in the bay, and when I stopped to check my hunting buddy, as I was getting off the snowmobile, I looked at my left and there was something weird sticking out of the ocean, on the ice. And I told my hunting buddy, 'What is that sticking out of the ice?' and he did not know. And I told him go back on the *komatik* and go see." Sammy Kogvik's pivotal role in locating the HMS *Terror* reveals a fundamental difference in approach.

The response of the new Prime Minister and government also differed from the statements made after the 2014 discovery. In 2014, the discovery "was presented to the public as the lynchpin of Canada's historic sovereignty of the Arctic archipelago and of the Northwest Passage" as the ministers and experts from Parks Canada to the Royal Canadian Geographical Society "all chimed in together from the same talking points on Franklin's foundational role in Canada's Arctic sovereignty" (Craciun 2016, 225). In 2016, not only was there no media or press conference with Prime Minister Trudeau announcing the find, but the chief spokesperson on the event was the Minister of the Environment, a portfolio that includes responsibility for Parks Canada (Van Dusen 2016). Saying that she was "thrilled about the discovery of HMS *Terror*" and "just as committed to working with the government of Nunavut and Inuit partners to protect and present all of the Franklin artifacts," Environment Minister Catherine McKenna emphasized Inuit involvement, stating that "Joint ownership of the artifacts from HMS *Erebus* with the Inuit Heritage Trust sets the stage for us to tell the stories of Nunavut's history, culture and heritage" (*Government of Canada Confirms Wreck*).[2] In the same press release, Nunavut's Minister of Culture and Heritage, George Kuksuk, expressed the territory's interest in working with the federal government, a process that is still ongoing. In 2016, the role played by Inuit in the discovery was highlighted in the media as well. In *Nunatsiaq News*, Louie Kamookak regretted that Parks had not taken local historians and elders seriously before: "There's a lot of modern information of a ship being seen there, under the water, from hunters and also from airplanes" (Ducharme).

Before the HMS *Erebus* was found, John Geiger wrote in the *Globe and Mail* that "much" is already known about the expedition from documents left by the explorers and from Inuit oral tradition. "Much has been surmised," writes Dorothy Harley Eber in *Encounters on the Passage*, "but it is the Inuit who have told most of what is known of the expedition's tragic history" (73). Indeed, David Woodman gathered together the documented Inuit testimony in *Unravelling the Franklin Mystery* 25 years ago, and Eber has traced it in oral tradition up to the present.[3] John Rae heard the first Inuit testimony in 1854 (Eber 70), and Knud Rasmussen recorded stories that not only described the fate of the expedition, but also revealed their important place in Inuit oral history. In *Fatal Passage: The Untold Story of John Rae, The Arctic Adventurer who Discovered the Fate of Franklin*, Ken McGoogan relies on historian Louis Kamookak to guide his party to the site where Rae erected a cairn marking the last link in the Northwest Passage, now called Rae Strait, where they in turn mounted a plaque commemorating John Rae's "discovery" (303–312).

Despite this long history of Inuit oral tradition, little notice of Inuit communities had been taken in the official announcements made in 2014. In their published account of the 2014 discovery, John Geiger and co-author Alanna Mitchell were careful to credit the Inuit, but in doing so, seemed to claim them for the nation: "Canadian scientists following their fellow Canadians' Inuit wisdom about where the ships sank" and noted that "the Canadian federal minister whose government department was in charge of the search was Leona Agglukkaq, an Inuk who comes from Gjoa Haven" (2015, 188). As sincere as this retrospective credit was, it contrasts reports of Inuit dissatisfaction. As Chris Sorenson reported in *Macleans*, residents of Gjoa Haven had not been pleased that the discovery of the HMS *Erebus* had been announced at "a tightly controlled press conference in Ottawa, nearly 3000 km away":

> In fact, the whole 2014 expedition at times smacked of style over substance. It involved uniformed members of the Royal Canadian Geographical Society, an expedition cruise ship full of dignitaries and a distracting visit from then prime minister Stephen Harper. (Sorensen)

Sorensen called the 2016 find "potentially bigger and more illuminating than the discovery of Erebus two years ago," but went on to observe that the truly impressive aspect of the recent discovery was the methodology used: "the ship's watery grave was pinpointed not by the latest technology or archeological theories, but thanks to [Adrian] Schimnowski's years-long efforts to build inroads with northern communities" (Sorensen). Schimnowski drew praise for listening to local people and especially for following up on the story told by the crew member Sammy Kogvik. When Kogvik recalled the incident that he describes in the video to Schimnowski, it was decided that the *Martin Bergman* should chart a course for the spot. "I trust him," Kogvik says. "But I don't trust Parks Canada because they like to keep everything secret" (Sorensen). Many of the articles covering the story commented on the role played by Inuit knowledge, particularly the story told by Sammy Kogvik to members of the Arctic Research Foundation (ARF) in the four-minute video of underwater footage of the wreck immediately posted by ARF.[4] The approach taken by ARF as it searched Arctic waters for the shipwreck appears to have taken the Indigenous knowledge carried in the stories of local people seriously.

Although some had latched on to the search for the ships "investing it with national purpose," Geiger reminded readers that the Inuit saw dead bodies onboard the ships, meaning "any discovery is likely to interest the British government as it would represent a burial ground" (*Ottawa to Mount Search*). Qaqortingeq told Rasmussen that his ancestors had seen dead bodies when they entered the ships:

> At first they were afraid to go down into the lower part of the ship, but after a while they grew bolder, and ventured also into the houses underneath. Here they found many dead men, lying in the sleeping places there; all dead. And at last they went down also into a great dark space in the middle of the ship. It was quite dark down there and they could not see. But they soon found tools and set to work and cut a window in the side. But here those foolish ones, knowing nothing of the white men's things, cut a hole in the side of the ship below the water line, so that the water came pouring in, and the ship sank. It sank to the bottom with all the costly things; nearly all that they had found was lost again at once. (Rasmussen 1999, 240)[5]

Discussion of their drowned remains recalls Geiger's work with Owen Beattie on *Frozen in Time* and turns attention from spoils and relics to sound a theme of loss more appropriate to the form of elegy found in Canadian literary culture. In George Elliott Clarke's elegy "*Lear of Whylah Falls*," for example, the speaker consoles the community mourning a young African-Nova Scotian man murdered after being falsely accused of speaking to a white woman, and the speaker, a member of the community, exhorts the people to "Let Othello sleep now" and to "mourn for all humanity" (86). Clarke's drowned hero calls to mind the drowned poet in A. M. Klein's "*Portrait of the Poet as Landscape*" or in Leonard Cohen's

"*Elegy*," as well as the many other drowned figures that Margaret Atwood observes in her Clarendon lectures published as *Strange Things: The Malevolent North in Canadian Literature*, including Gwendolyn MacEwen's sunken *Terror and Erebus* in the verse play by that name, but it is Clarke's allusion to John Milton's *Lycidas* that is a touchstone here. Clarke's Othello goes "To sound the wrinkling and remorseless deep/ That shut over the head of Lycidas" the drowned poet in Milton's elegy. The speaker in that poem consoles the mourners with the news:

> For Lycidas your sorrow is not dead,
> Sunk though he be beneath the wat'ry floor:
> So sinks the day-star in the ocean bed
> And yet anon repairs his drooping head,
> And tricks his beams, and with new-spangled ore
> Flames in the forehead of the morning sky:
> So Lycidas sunk low, but mounted high... (Milton 124–125)

The lost one is transformed into the "genius" or "guardian spirit," an eternal presence presiding over the shores of his own demise. In cultural terms, transformation into a presiding spirit was the fate of Sir John Franklin "sunk low, but mounted high" in the tradition of Arctic narrative. At the time of Franklin's disappearance, as Russell Potter explains, "a paradigmatic shift occurred in late Victorian British imagination, as the nation's patriotic feelings seem to have been fuelled less by the sublimity of sacrifice than by a sense of loss and mourning" (Potter 2007, 2–3). As Margaret Atwood puts it: "Because Franklin was never really 'found,' he continues to live on as a haunting presence; certainly in Canadian literature" (1995, 16). The Franklin expedition initiated further searching from the British admiralty's efforts at the time to the present day with Lady Jane Franklin's determination to keep the search alive is as much a part of the expedition's legacy as the voyage itself.

By reminding readers that the sunken vessels cradle human remains, Geiger drew on the Inuit accounts that were vilified in Victorian England and that still often receive skeptical treatment today. The lingering skepticism toward Inuit accounts is documented in John Walker's *Passage*, a film that explores the British response to John Rae's findings in the Arctic. Based on Ken McGoogan's *Fatal Passage*, the film works to create the illusion that it is what Darrell Varga calls "a process of discovery" rather than "a set of fixed ideas" (Varga 52). It achieves this effect through recognizable metafictional techniques such as interlacing dramatic scenes with footage of the director, actors, and advisors as they workshop the script (Varga 52). The film follows John Rae, sensitively portrayed by Rick Roberts, from his home in the Orkneys to the Arctic and back to the room in Admiralty House, London where his report of Inuit testimony to the last days of the Franklin expedition was first received. In it, Rae had reported accounts from Inuit in the area which included evidence that some members of the expedition had engaged in cannibalism before they died.

When John Rae submitted his account of Inuit testimony, his contemporaries were loath to imagine British naval officers and seamen as cannibals, nor were they willing to entertain the idea of failure on the part of the expedition. As Atwood puts it, "the best-equipped ships of their age, offering the latest in technological advances" had failed and "defeat of such magnitude called for denial of equal magnitude" (2014, 5). Rather than acknowledge the fate of Franklin, the British made monuments to him as the discoverer of the Northwest Passage, and the Inuit accounts were thus discredited in the press, notably by Charles Dickens in an article in *Household Words*. In the film, Walker brings Inuit elder Tagak Curley to see one such monument in London. When Walker points out that a figure in the frieze depicting Franklin's funeral in the Arctic is holding a bow and arrow, Curley notes that the expedition failed because in reality, there were no Indigenous people with the expedition. It was "a fatal mistake" because as Curley says, "We could have helped them a lot." Next, they go to Admiralty House where they meet Ken McGoogan and sit down with a number of experts including retired British naval officer and historian Lieutenant Ernest Coleman. Basing his conclusions on his own analysis of forensic evidence, Coleman disbelieves the Inuit testimony that Rae and others collected and rejects the claim that members of the expedition engaged in cannibalism; rather, he adheres to the view that knife marks found on their remains indicate defensive wounds received while under attack by the Inuit. This view arose soon after the expedition disappeared and appears in the article in which Dickens disputed Rae's findings mainly on the grounds that the British did not have a history of cannibalism. Dickens dismissed the testimony of Inuit as "kettle-stories" that may have been embellished or distorted in translation. As to the trustworthiness of the Inuit informants, Dickens deemed them to be of the same character as "savages" everywhere:

We believe every savage to be in his heart covetous, treacherous, and cruel; and we have yet to learn what knowledge the white man—lost, houseless, shipless, apparently forgotten by his race, plainly famine-stricken, weak, frozen, helpless, and dying—has of the gentleness of Esquimaux nature. (Dickens, 363)

For Tagak Curley, this version of events is an insult and affront to the Inuit who are depicted not only as liars but as the murderers of weak and starving men, a slander for which Curley says, "Someone really ought to apologize." Like his Victorian predecessors, Coleman is unable to believe that British seaman practiced cannibalism even though others corroborated Rae's report at the time (Beattie and Geiger 114–115). Walker invites Gerald Dickens, a descendent of the novelist, who is clearly embarrassed to hear about his ancestor's animosity toward the Inuit. He persists in claiming it was "out of character" until Ken McGoogan reads a passage in which Charles Dickens advocates extermination of those involved in the Indian Mutiny. On behalf of his family, the visibly chastened Dickens offers an apology to Tagak Curley and his people which Curley graciously accepts. For Darrell Varga, "This scene exemplifies the resilient power of documentary to bring history into the present in ways that could not be expected but that then have concrete implications in the world" (100). While the moment of reconciliation between the two men is both touching and significant, its place in the film is clearly meant to highlight Walker's achievement for including "what has been left out of the story," that is, the voice of the Inuit (Varga 52). For Russell Potter, however, the film's staging of this moment of reconciliation is somewhat less impressive as it is unlikely to have any lasting effect on the people of Nunavut (2016, 97). It is also questionable whether the film alters the course of Arctic representation simply by claiming that this well-known story is "untold." After all, Canadian writers have been telling the truth that Victorian Britain did not wish to hear about Arctic exploration for a long time. In Rudy Wiebe's novel of Franklin's 1819 expedition, *A Discovery of Strangers*, for example, one of the characters, an ordinary seaman, shows the extremes to which British sailors were sometimes forced to go, saying that "any English tar" (286) will tell you "Meat … is meat … An' any officer'll tell you the same, if he don't lie" (293). It is because the stories are so well known that they become part of literary history. Wiebe was writing after Owen Beattie's team concluded that they had found "physical evidence to support Inuit tales of cannibalism among the dying crewmen" and added their findings to the extensive body of Arctic archeology on the subject (Beattie and Geiger 110). Knowing it to be a proven fact that some of the crew cannibalized others, Margaret Atwood spoofs British disdain in *Strange Things* as she wonders how to make her first lecture more engaging:

> The English, I knew, were very fond of cannibalism. If I could put some of that in, I was off on the right foot. And so it turned out; at the sherry party after the first lecture, I was treated to the spectacle of a number of Oxford academics nibbling hors-d'oeuvres and delicately discussing the question of who they would be prepared to eat. (1995, 2–3)

As Atwood observes, "each age has created a Franklin suitable to its needs" (2014, 4), and her wry humor suits our age of irony. Unfortunately, media coverage of the ships' discovery did not acknowledge the long and rich literary afterlife the Franklin expedition has enjoyed. If it had, it would have told a different story, not one of heroic discovery. In Canadian literature, searching for Franklin searching for the Northwest Passage has served as a metaphor for a search for meaning that never ends. The repetition of the Franklin story has transformed it from historical narrative to a myth, then demythologized it, and each retelling infuses the story with new meanings that need to be read in their specific contexts.[6] Even today, as Craciun argues, "the search for Franklin, and the sport of not finding Franklin, has centered around a productive absence with so much dynamic significance that it eventually propelled the Franklin disaster back into the sphere of international geopolitics" with the Canadian government's sponsorship of searches beginning in 2011 (Craciun 2016, 81). For Craciun, "the idea of the Franklin disaster relics as archived and awaiting rediscovery continues to play a significant role in a wide range of claims made upon, and voyages made in, the Arctic" (2016, 81). In Canadian literature, the Franklin expedition has been thoroughly dissected, demystified, and then remythologized as a cautionary tale of overreaching ambition: from Atwood's "The Age of Lead" and Mordecai Richler's *Solomon Gursky Was Here* to poems by Al Purdy and Gwendolyn MacEwen and many others.[7] And this is only to name works of "literary" fiction. More can be found in Inuit oral history, in academic works and popular culture, like Stan Rogers' song "Northwest Passage." Ironically, volumes have been generated in the "productive absence" of Franklin and his ships (see Craciun 2016), and these literary explorations reflect how the Northwest Passage has been the site of differing worldviews where the limits of knowledge are tested. Not surprisingly, literary knowledge is not always given a place in representations like "Great Canadian North" that seek to paper over differences in the promotion of Canadian sovereignty. As Craciun argues in *Writing Arctic Disaster*, the "continuing fascination" with the Franklin expedition can be "a distraction" (232), taking our attention from the real threats to finding a better future for the Arctic.

One of the first literary works inspired by the Franklin expedition, Gwendolyn MacEwen's verse play *Terror and Erebus*, reinterprets the Franklin expedition as what historians now agree was the "combination of hubris, poor preparation and technological inadequacies, endemic to the Admiralty's Eurocentric approach to exploration" (Craciun 2012, 3). *Terror and Erebus* is one of the many literary revisions of the Franklin story that undermines this myth of discovery and exploration, shifting the way we think about Arctic history in Canada; yet MacEwen's play is not merely corroboration of modern consensus on the Franklin expedition, nor is it a gloss on the cultural history of Victorian Britain, another post-colonial voice speaking back to an arrogant imperial center. Fifty years before the ships would be found; before David Woodman's *Unravelling the Franklin Mystery* collected the stories recorded by Rasmussen and others; before the modern oral history compiled in Dorothy Eber's *Encounters on the Passage*, or agencies like Parks Canada and Arctic Research Foundation would make consulting local people part of its investigation, MacEwen revised the Franklin story and placed Indigenous knowledge at the center of literary experience. Gwendolyn MacEwen knew that the Inuit witnesses knew the truth; several truths, about Franklin and his gallant crew (see Hulan 2015). In *Terror and Erebus*, the character 'Rasmussen' establishes Indigenous knowledge of the environment as an alternative way of knowing available to the white men in the Franklin expedition but whose cultural blindness prevents them from seeing. He despairs at Franklin's failure to adopt the Inuit knowledge and technology, the "wooden slits" that would have "tamed the light" and helped them see. Snow blindness serves as a metaphor connecting the challenge faced by European science and technology in the unknown environment to the expedition's moral and spiritual crisis. As the men of the expedition scattered these instruments in the snow, thinking they have come to "the end of science," Rasmussen knows that even this recognition of their failure is misguided. MacEwen's Franklin expedition serves to illustrate the limits of European scientific knowledge and technology faced with the Arctic landscape, a critique that clearly had no impact on the creators of the "Great Canadian North."

MacEwen's fellow poet and friend Margaret Atwood would also explore the devastation caused by Modernity by conjuring the Franklin expedition. At the end of Atwood's short story "The Age of Lead," the narrator offers this sinister look at the ordinary things in the protagonist Jane's kitchen:

> Her toaster oven, so perfect for solo dining, her microwave for the vegetables, her espresso maker—they're sitting around waiting for her departure, for this evening or forever, in order to assume their final, real appearances of purposeless objects adrift in the physical world. They might as well be pieces of an exploded spaceship orbiting the moon. (1991, 174–175)

People are dying of cancer, heart attacks, and mysterious viruses, Jane has noticed, and though she longs to move out of the polluted city, she also realizes there is nowhere to go, no way to escape modern life. Like her namesake, Jane is on the search for Sir John Franklin though by watching a television documentary that presents the theory that the Franklin expedition succumbed to lead poisoning from the solder used in the cans of food they brought with them. After the exhumation of the bodies and tests on them, the scientists conclude that: "It was what they'd been eating that had killed them" (174). Scenes from the documentary, John Torrington's "tea-stained eyes" (160), his bare feet and flannel trousers (167), so recognizable to anyone who has seen the photographs in Owen Beattie and John Geiger's *Frozen in Time*, are scattered through a narrative of Jane's friendship with Vincent who has recently died from "a mutated virus that didn't even have a name yet" (173). The story ends with an ambivalent vision:

> Increasingly the sidewalk that runs past her house is cluttered with plastic drinking cups, crumpled soft-drink cans, used take-out plates. She picks them up, clears them away, but they appear again overnight, like a trail left by an army on the march or by fleeing residents of a city under bombardment, discarding objects that were once thought essential but are now too heavy to carry. (175)

This final scene recalls how Franklin's men scattered objects as they tried to move south leaving trails "like the trails in fairy tales, of bread crumbs or seeds or white stones" (161). Franklin's discarded things became collectible relics of the expedition carefully rendered in *The London Illustrated*

News and put on display in Greenwich. In Atwood's story, all the debris of consumer society seems like Franklin's books and silver spoons jettisoned for the sake of survival. Those imaginary hordes fleeing the city are stripping down to the essentials but, like their Victorian counterparts, cannot seem to do without material goods. By ending with the statement that these things are "now too heavy to carry," the narrator comments on the main theme, leaving the reader to consider the costly consequences of Modernity.

AFTER NORDICITY

For the authors I studied in *Northern Experience and the Myths of Canadian Culture*, ideas of the North, whether of the Arctic itself or its adjacent regions, were shaped by a sense of permanence. The imagined North was an eternal, unchanging physical and imaginary landscape of vast white spaces, landfast ice, permafrost, and a once impassable Northwest Passage. In the late twentieth century, Canadian authors showed a passionate enthusiasm for this imagined North and encouraged fellow citizens to grasp their unique northern heritage. The gap between the "real" and the "imagined" North had been a theme in Canadian writing for some time, yet the growing awareness that people living in the North have a distinct perspective had not translated into greater authority for northern voices. In *Northern Experience*, I argued that the Canadian national identity constructed from this imagined North served to justify the colonial relationship between the Canadian state and real people living in the northern regions. If awareness of the North as a place with its own reality has grown in recent decades, this consciousness has not always displaced enduring literary myths of the North as a frontier, nor has it prevented a resurgence of nationalist rhetoric in policies on resource development and territorial sovereignty reflected in the search for Franklin's ships. The literary exploration of the North's intrinsic value generated much discussion and many texts, but it too did not lead to a change. The colonial relationship between Canada and northern regions persisted and still persists today. It was my contention that the writing by outsiders and travelers could do little to alter the situation if it continued in the nationalist mode. Although most of the writers were explicit in their desire to bring awareness to the North and its people, the unintended consequence was a bolstered sense of Canadian ownership.

"It's Your North, Too" the subtitle of Louis-Édmond Hamelin's monumental study *Canadian Nordicity* announced as many of Canada's poets and writers were being drawn northward. *Nordicity* continues to resonate in Canadian and Québécois cultures as an idea conceived out of the desire to understand the environments in which we live. Ironically, the term *nordicité*, which Canadian nationalists embraced in translation, was one of the intellectual innovations of *la révolution tranquille* that contributed to the movement for Quebec sovereignty that threatens the federation Canadian nationalists revere. As Daniel Chartier observes, *nordicité* "est même devenu pour les Québécois l'un des principaux mots pour rendre compte de leur identité" (6). Hamelin's theory, although deeply grounded in empirical data and observation, can also be seen as a forerunner to the theories of the relational production of space advanced in the late twentieth century. With its gradations of northern experience, Hamelin's system lent itself to application in the social and physical sciences, and it established the relational nature of northern places. In Canadian literary history, the Arctic was also imagined as a distinct part of "the North," a relational term that depends on the observer's location and yet is often imagined as a space that encompasses all of Canada (see Hulan 2002). There are similarities in the history of Arctic and sub-Arctic regions that may support this generalized definition; however, it serves mainly to confer northern identity on those living far from the Arctic and its peoples. Arctic studies in Canada have focused on policy and politics around international relations, resource extraction, trade, and human resources while the GOC has mounted arguments that push Canadian boundaries out to the edge of the ice used by Inuit hunters, and claim sovereignty over the Northwest Passage by citing the history of exploration and instances of recent navigation through it.[8] Indeed, the idea of "Canada as North" relies on the contiguity of the Arctic. But Canada is not the only Arctic nation, and Arctic studies is as vast a field as the region for which it is named, as varied as the perspectives of the diverse cultures and nations in it: it is often viewed set apart from other places with its own characteristics, and it has served as the object of scientific investigation from every discipline. Perhaps for this reason, the study of the Arctic is a site of contestation in the debate between scientific and Indigenous forms of knowledge. While governments and industry look for ways to exploit the natural resources in the Arctic, often capitalizing on the warming temperatures to do so, the consequences of these activities are coming home.

In *Cold Matters: Cultural Perceptions of Cold, Ice and Snow,* Heidi Hansson and Catherine Norberg remark the shift in the literary discourse associated with polar regions affected by global warming. Historically, they note, "ice" signifies beauty and "snow" purity, but "cold" is most often associated with death. In literature, "the Arctic and the Antarctic have often been perceived as masculine-coded areas" (Hansson and Norberg 7) where men pit themselves against harsh, dangerous conditions created by cold, ice and snow. The stereotyped struggle between "man" and "Nature" is found at the core of many literary representations in which "[w]inter is overrepresented" and "the summer season seems almost not to exist" (Hansson and Norberg 8). The depiction of Arctic and Antarctic regions as "harsh, frightening and potentially deadly" (Hansson and Norberg 8) and the association of these forbidding elements as feminine is explored by Margaret Atwood in *Strange Things: The Malevolent North in Canadian Literature.* These retrospective analyses of literary representation, including my own work, aim to increase our understanding of the gendered and racialized narratives dominating literature in the nineteenth and twentieth centuries, and many of the tropes analyzed within them continue to circulate in the present. Even though the stereotyped representation of the "Great White North" reflected in images of the North as a vast unchanging whiteness in the Canadian imagination has been thoroughly analyzed and critiqued, this racialized imaginary is still evoked in nationalist arguments (see Baldwin et al.). At the same time, Hansson and Norberg acknowledge that the threat posed by global warming has generated new narratives about the Arctic, including narratives that may have appeared counterintuitive not long ago:

> At a time when the survival of cold regions is threatened, it is vital to change the paradigm that figures cold as negative and instead highlight its positive characteristics. Apart from emphasizing the necessity of cold matters, such a paradigm change could have radical implications for all the symbolic and metaphorical uses of cold. Instead of routinely associating cold with death, it is essential to show its crucial importance for continued life. (Hansson and Norberg 21)

By echoing the title of *History Matters,* feminist historian Gerda Lerner's call to heal the wounds of history by uncovering the 'inconvenient' truths of the past, Hansson and Norberg seek to reconceptualize the Arctic in light of the inconvenient truths of the present.

Literary representations of the Arctic have changed since I began writing about the Canadian North twenty-five years ago. With the growing awareness of the effects of climate change, writers are using the framework of national identity to both articulate and deconstruct the myths and stories that create the idea of Canada as North. At the end of Margaret Atwood's *Strange Things*, the discussion takes this turn:

> The edifice of Northern imagery we've been discussing in these lectures was erected on a reality; if that reality ceases to exist, the imagery, too, will cease to have any resonance or meaning, except as a sort of indecipherable hieroglyphic. The North will be neither female nor male, neither fearful nor health-giving, because it will be dead. The earth, like trees, dies from the top down. The things that are killing the North will kill, if left unchecked, everything else. (116)

With this conclusion, Atwood puts an end to the discussion of a dichotomy between the real and the imagined North by acknowledging the fact that, however remote they may have been, however fanciful and fictional, the images found in Canadian literature and culture refer to a reality that exists and is lived.

As the ice melts, the Arctic is increasingly represented as suffering the excesses of our wasteful, industrial Modern way of life; however, the exploration and exploitation of the Arctic has been central to the industrial and technological developments that constitute Modernity. The convergence of economic development and national security interests has always motivated government policy and activity, but these interests have taken on greater urgency as global warming opens the Northwest Passage creating greater access to resources such as oil, gas, and diamonds. The Arctic created in the literary imagination is being gradually reshaped by the current state of the Arctic environment, and with it, ideas of Canadian identity and culture. As the Northwest Passage remains open longer each year, more and more visitors are traveling in the Arctic adjacent to Canada. While this increased activity has promoted awareness of how climate change is affecting the Arctic, it is also having an impact on the environment and on the people who rely on it. By drawing attention to what Canadian readers share in common with other citizens of the circumpolar and wider world, the recent works by Kathleen Winter and Sheila Watt-Cloutier studied here signal a shift in Canadian writing from a focus on national identity and Canadian sovereignty to a circumpolar and global outlook influenced by the voices of the sovereign Indigenous peoples of the Arctic.

NOTES

1. Varieties of this legend are heard across the Arctic. In *This Is My Country*, Noah Richler reproduces his transcription of Elder George Kappianaq's version of Uinigumasuittuq (Toronto: McClelland, 2006, 91–92).
2. Russell A. Potter's blog follows these events and their coverage in the media (visionsnorth.blogspot.com).
3. In 1854, John Rae heard the first Inuit testimony concerning Franklin. Motifs threading through these stories include encounters with white men, some of them peaceful and others violent, ships that are frozen in the ice, and others that sink. While Inuit storytellers are motivated by many different objectives in telling these stories, as Eber recognizes, non-Inuit are often most curious about what happened to Franklin's ships *Terror* and *Erebus*. Writing before the location of the *HMS Erebus* in 2014 and the *Terror* in 2016, Eber collects a number of accounts of ship sightings and sinkings around King William Island. In 1909, J. B. Tyrell published the statements of Thomas Mustagan and Paulet Panakies who reported seeing ships off King William Island in 1890. The story of sinking ships is "still alive in Gjoa Haven" where Eber recorded it in 1999. One of her informants, Matthew Tiringaneak, tells of people ripping up papers on Chantury Island just as Qaqortineq told Rasmussen.
4. A shortened version of the video accompanies the *Macleans* article (www.macleans.ca/news/canada/how-trust-led-to-hms-terror/).
5. See also Rasmussen's *The Netsilik Eskimos* (New York: AMS, 1976, 130), and David C. Woodman, *Unravelling the Franklin Mystery: Inuit Testimony* (McGill-Queens University Press, 1991, 218–219).
6. The Franklin expedition is one of the most often represented episodes in Canadian history and literature. Sherrill E. Grace observes, "Franklin lives in the Canadian imagination, in the work of playwrights, poets, novelists, filmmakers, and scholars who are drawn to participate in a myth of the North" and "he lives in our continuing, insistent demand for answers, our rapacious need to uncover and possess, our transgressive, objectifying gaze, our continued search for the Northwest Passage to origins and truth," ("'Franklin Lives': Atwood's Northern Ghosts," *Various Atwoods*. Ed. Lorraine York. Toronto: Anansi, 1995: 162–163).
7. The Franklin expedition left a documentary and cultural trail that many in Canadian Studies have followed, including I. S. MacLaren, John Moss, Sherrill Grace, and David Woodman among others. See also Scott Cookman's *Ice Blink* (New York: Wiley, 2000) and Adriana Craciun's *Writing Arctic Disaster* (Cambridge University Press, 2016).
8. For comprehensive studies of the issues surrounding Canadian sovereignty in the Arctic, see Shelagh Grant's *Polar Imperative: A History of Arctic Sovereignty in North America* (Vancouver: Douglas and McIntyre, 2010);

Michael Byers' *Who Owns the Arctic?* (2009); and *Canada and the Changing Arctic* (Vancouver: Douglas and McIntyre, 2011) by Griffiths, Huebert, and Lackenbauer (Waterloo: Wilfrid Laurier University Press, 2011).

REFERENCES

Atwood, Margaret. 1991. The Age of Lead. In *Wilderness Tips*, 159–175. Toronto: McClelland.
———. 1995. *Strange Things: The Malevolent North in Canadian Literature.* Oxford: Clarendon.
———. 2014. Introduction. In *Frozen in Time: The Fate of the Franklin Expedition*, 1–8. Vancouver: Greystone.
Baldwin, Andrew, Laura Cameron, and Audrey Kobayashi, eds. 2011. *Rethinking the Great White North: Race, Nature, and the Historical Geographies of Whiteness in Canada.* Vancouver: University of British Columbia Press.
Beattie, Owen, and John Geiger. 2014 [1988]. *Frozen in Time: The Fate of the Franklin Expedition.* Vancouver: Greystone.
Berger, Thomas R. 1988. *Northern Frontier, Northern Homeland.* Rev. ed., 2 vols. Toronto: Douglas and McIntyre.
Byers, Michael. 2009. *Who Owns the Arctic? Understanding Sovereignty Disputes in the North.* Vancouver: Douglas and McIntyre.
Cameron, Emilie. 2015. *Far Off Metal River: Inuit Lands, Settler Stories, and the Making of the Contemporary Arctic.* Vancouver: University of British Columbia Press.
Canada Launches New Arctic Search for Franklin's Lost Ships. 2008. *CBC*, 15 August. Accessed September 5, 2008. www.cbc.ca/news/canada/north/canada-launches-new-arctic-search-for-franklin-s-lost ships-1.702857
Chartier, Daniel, and Jean Désy. 2014. *La Nordicité du Québec: Entretiens avec Louis-Edmond Hamelin.* Québec: Presses de l'Université du Québec.
Clarke, George Elliott. 2010 [1990]. Lear of Whylah Falls. *Whylah Falls*, 86. Rev. 3rd ed. Kentville, NS: Gaspereau.
Coates, Kenneth. 1985. *Canada's Colonies: A History of the Yukon and Northwest Territories.* Toronto: James Lorimer.
Coates, Kenneth, and Judith Powell. 1989. *The Modern North: People, Politics and the Rejection of Colonialism.* Toronto: James Lorimer.
Cookman, Scott. 2000. *Ice Blink: The Tragic Fate of Sir John Franklin's Lost Polar Expedition.* New York: Wiley.
Craciun, Adriana. 2012. The Scramble for Franklin's Grave. *Literary Review of Canada* 20 (4): 3–5.
———. 2016. *Writing Arctic Disaster: Authorship and Exploration.* Cambridge: Cambridge University Press.
Dickens, Charles. 1854. The Lost Voyagers. *Household Words* 2: 362–365.

Ducharme, Steve. 2016. HMS Terror, Franklin's Second Ship Finally Found in Nunavut. *Nunatsiaq News Online*, 12 September. Accessed September 15, 2016.www.nunatsiaqonline.ca/stories/article/65674hms_terror_franklins_second_ship_finally_found_in_nunavut/

Eber, Dorothy Harley. 2008. *Encounters on the Passage: Inuit Meet the Explorers*. Toronto: University of Toronto Press.

Edward, Olivia. 2010. The Search for Franklin. *Geographical* 82 (2): 30–33.

Fee, Margery. 2015. *Literary Land Claims: The "Indian Land Question" from Pontiac's War to Attawapiskat*. Waterloo, ON: Wilfrid Laurier University Press.

Freeman, Mini Aodla. 2015. *Life Among the Qalunaat*. Edited by Keavy Martin, Julie Rak, and Norma Dunning. Winnipeg, MB: University of Manitoba Press.

Geiger, John. 2008. Ottawa to Mount Search for Lost Ships. *Globe and Mail*, 14 August. Accessed September 5, 2008. http://www.theglobeandmail.com/news/national/ottawa-to-mount-search-for-lost-franklin-ships/article658085/

Geiger, John, and Alanna Mitchell. 2015. *Franklin's Lost Ship: The Historic Discovery of the HMS Erebus*. Toronto: HarperCollins.

Government of Canada Confirms Wreck of HMS Terror and Deepens Collaboration with Inuit in Nunavut Through Joint Ownership of Franklin Artifacts. 2016. *Canadian Newswire*, 26 September. Accessed October 1, 2016. www.newswire.ca/news-releases/government-of-canada-confirms-wreck-of-hms-terror-and-deepens-collaboration-with-inuit-in-nunavut-through-joint-ownership-of-franklin-artifacts-5

Grace, Sherrill E. 1995. 'Franklin Lives': Atwood's Northern Ghosts. In *Various Atwoods: Essays on the Later Poems, Short Fiction, and Novels*, ed. Lorraine York, 146–166. Anansi: Toronto.

———. 2001. *Canada and the Idea of North*. Montreal: McGill-Queens University Press.

Grant, Shelagh. 1988. *Sovereignty or Security: Government Policy in the Canadian North, 1936–1950*. Vancouver: University of British Columbia Press.

Grant, S.D. 1989. Myths of the North in the Canadian Ethos. *Northern Review* (3/4): 15–41.

Grant, Shelagh. 2010. *Polar Imperative: A History of Arctic Sovereignty in North America*. Vancouver: Douglas and McIntyre.

Great Canadian North. 2015. Video. Accessed May 20, 2015. www.Canada.pch.gc.ca/eng/14353195587305

Griffiths, Franklyn. 1987. *The Politics of the Northwest Passage*. Montreal: McGill-Queens University Press.

Griffiths, Franklyn, Rob Huebert, and P. Whitney Lackenbauer. 2011. *Canada and the Changing Arctic: Sovereignty, Security, and Stewardship*. Waterloo, ON: Wilfrid Laurier University Press.

Hamelin, Louis-Édmond. 1978. *Canadian Nordicity: It's Your North, Too*. Translated by William Barr. Montreal: Harvest.

Hansson, Heidi, and Cathrine Norberg, eds. 2009. *Cold Matters: Cultural Perceptions of Snow, Ice and Cold*. Northern Studies Monographs no. 1, University of Umeå and the Royal Skyttean Society, Umeå, Sweden.

Hodgins, Bruce W. 1977. *The Canadian North: Source of Wealth or Vanishing Heritage?* Scarborough: Prentice-Hall.

Hulan, Renée. 2002. *Northern Experience and the Myths of Canadian Culture*. Montreal: McGill-Queens University Press.

———. 2015. *Terror and Erebus* by Gwendolyn MacEwen: White Technologies and the End of Science. *Nordlit* 35: 122–135.

MacEwen, Gwendolyn. 1988. *Terror and Erebus*. In *Afterworlds*, 41–57. Toronto: McClelland.

Martin, Keavy. 2012. *Stories in a New Skin: Approaches to Inuit Literature*. Winnipeg, MB: University of Manitoba Press.

McGoogan, Ken. 2001. *Fatal Passage: The Untold Story of John Rae, The Arctic Adventurer who Discovered the Fate of Franklin*. Toronto: HarperCollins.

Milton, John. 1957 [1637]. Lycidas. In *John Milton: Complete Poems and Major Prose*, ed. Merritt Y. Hughes, 120–125. New York: Macmillan.

Nappaaluk, Mitiarjuk. 2014 [1983]. *Sanaaq: An Inuit Novel*. Translated by Bernard Saladin D'Anglure. Translated from French by Peter Frost. Winnipeg, MB: University of Manitoba Press.

Nuliajuk: Mother of the Sea Beasts. 2001. Dir. John Houston and Peter d'Entremont. Triad Films.

Nungak, Zebedee, and Eugene Arima, eds. 1969. *Unikkaatuat sanaugarngnik atyingnaliit Puvirgni turngmit/Eskimo Stories from Povungniyuk, Quebec*. Ottawa: Queen's Printer.

Passage. 2008. Dir. John Walker. National Film Board of Canada.

Petrone, Penny, ed. 1988. *Northern Voices: Inuit Writing in English*. Toronto: University of Toronto Press.

Potter, Russell. 2007. *Arctic Spectacles: The Frozen North in Visual Culture, 1818–1875*. Montreal: McGill-Queens University Press.

———. 2016. *Finding Franklin: The Untold Story of a 165-year Search*. Montreal: McGill-Queens University Press.

Qalunaat! Why White People are Funny. 2006. Mark Sandiford in collaboration with Zebedee Nungak. NFB.

Rasmussen, Knud. 1976. *The Netsilik Eskimo: Social Life and Spiritual Culture*. New York: AMS.

———. 1999 [1927]. *Across Arctic America: Narrative of the Fifth Thule Expedition*. Fairbanks, AK: University of Alaska Press.

Richler, Noah. 2006. *This Is My Country, What's Yours? A Literary Atlas of Canada*. Toronto: McClelland.

Sangster, Joan. 2016. *The Iconic North: Cultural Constructions of Aboriginal Life in Postwar Canada*. Vancouver: University of British Columbia Press.

Ship Found in Arctic 168 Years after Doomed Northwest Passage Attempt. 2016. *The Guardian*, 12 September. Accessed September 13, 2016. www.the-guardian.com/world/2016/sep/12/hms-terror-wreck-found-arctic-nearly-170-years-northwest-passage-attempt/

Sorensen, Chris. 2016. How One Man's Efforts to Build Inroads with Northern Communities Helped Uncover the Second of Sir John Franklin's Doomed Ships. *Macleans*, 14 September.

Tyrell, J.B. 1908–1909. A Story of a Franklin Search Expedition. *Transactions of the Canadian Institute* 8: 393–402.

Van Dusen, John. 2016. Nunavut Shipwreck Confirmed as Sir John Franklin's HMS Terror. *CBC News*, 26 September. Accessed September 26, 2016. www.cbc.ca/news/canada/north/hms-terror-confirmed-1.3779127

Varga, Darrell. 2012. *John Walker's Passage*. Toronto: University of Toronto Press.

Watt-Cloutier, Sheila. 2015. *The Right to Be Cold: One Woman's Story of Protecting her Culture, the Arctic, and the Whole Planet*. Toronto: Allen Lane.

Wiebe, Rudy. 1994. *A Discovery of Strangers*. New York: Knopf.

Woodman, David C. 1991. *Unravelling the Franklin Mystery: Inuit Testimony*. Montreal: McGill-Queens University Press.

Wynn, Graeme. 2007. *Canada and Arctic North America: An Environmental History*. Santa Barbara, CA: ABC-CLIO.

Younger-Lewis, Greg. 2004. Armed Forces to Pour More Money into North. *Nunatsiaq News*, 23 July. www.nunatsiaqnewsonline.ca/stories/article/armed_forces_to_pour_more_money_into_north/

CHAPTER 2

Becoming *Boundless*: Kathleen Winter's Arctic Excursion

Abstract In a close reading of Kathleen Winter's memoir *Boundless*, Hulan examines the effect Arctic tourism, particularly the proliferation of Arctic cruises made possible by global warming, has had on writing about the Arctic. In 2010, Winter was writer-in-residence on the cruise ship *Clipper Adventurer* when it went aground in the Northwest Passage. During the excursion, she documents her growing awareness of Indigenous knowledge and her attempts to learn from Aaju Peter and Bernadette Dean, the Inuit women who act as cultural interpreters on the cruise. After her return, Winter continues a journey from tourist to witness to the legacy of colonialism. Because Winter sets out to understand the Arctic world without borders, *Boundless* offers an example of writing about the Arctic that is less preoccupied with Canadian national and territorial sovereignty.

Keywords Arctic cruise • Arctic tourism • *Clipper Adventurer* • Kathleen Winter

In the spring of 2016, just months before the location of the HMS *Terror*, the CBC reported that the *Crystal Serenity* would make a 32-day journey through the Northwest Passage in August 2016. An estimated 1000 passengers would pay between $30,000 and $156,000 per person to travel from Anchorage, Alaska, through the Northwest Passage stopping in

© The Author(s) 2018
R. Hulan, *Climate Change and Writing the Canadian Arctic*,
https://doi.org/10.1007/978-3-319-69329-3_2

communities in the Northwest Territories, Nunavut, and Greenland before eventually reaching New York City. According to the report, the Canadian Coast Guard would use the opportunity "to evaluate the risks, the challenges and be sure with Transport Canada if the regulations are appropriate" and to plan for future cruises in the area. As Michael Byers warned, the "Canadian Forces' search-and-rescue helicopters sometimes require two days to reach the Northwest Passage from their bases in Newfoundland, Nova Scotia and British Columbia," and an incident involving the *Crystal Serenity* could stretch the Canadian Coast Guard beyond its current capabilities because existing "search-and-rescue systems would be overwhelmed by any accident in the Arctic involving more than a few dozen people" (*Globe and Mail*, 18 April 2016). While the ship was expected to have navigational experts, radar, and iceberg sighting technology on board as well as an icebreaker traveling with it, the CBC sounded the alarm about potential dangers in the largely uncharted waters of the Northwest Passage, recalling the 2010 grounding of the *Clipper Adventurer*, a ship registered in the Bahamas but operated by Adventure Canada, headquartered in Mississauga, Ontario.[1]

The Transport Safety Board of Canada (TSB) found that the "*Clipper Adventurer* ran aground on a previously reported but uncharted shoal after the bridge team chose to navigate a route on an inadequately surveyed single line of soundings" ("Marine Investigation Report M10H0006"; see also George 2012). In essence, the crew had decided to take a shorter but uncharted route to Kugluktuk without checking available reports beforehand. The ship's forward looking sonar, which could have detected the shoal, was not working, and the ship was traveling at full speed, 7 knots over the 6 knots recommended for the route. It had been a kind of hubris that had led to this mistake, a hubris that so often characterizes Arctic disaster in fiction and reality. In their analysis of the incident, Emma Stewart and Jackie Dawson concluded that "[i]t was a matter of good fortune that the *Amundsen* was relatively close" to the scene "and even more fortunate that the icebreaker was carrying the appropriate equipment and operational expertise to map a safe rescue course" (265). Stewart and Dawson called for the creation of a common governing body, along the lines of those operating in Greenland and Norway, to oversee the cruise industry in Canada (266; see also Sandiford). So far, no such body exists. The Coast Guard sued the company for the expenses incurred by the rescue after Adventure Co. sued the federal government for failing to update naval charts of the region. Although the 120 passengers, ranging in age from 15 to 90, and 60 crew on board the *Clipper Adventurer* were rescued, the incident exposed the risks posed by modern polar adventure, both to travelers and to the Arctic environment. The people of Kugluktuk gave the stranded passengers a warm welcome, but the incident also drew attention to Canada's lack of emergency preparedness. *Nunatsiaq News* reported that, with no emergency plan in place and everyone in Kugluktuk literally "gone fishing," it was only by chance that the community was notified at all (George 2010). One member of the community "who helped out during the night in Kugluktuk—but asked not to be named—said that if Canada plans to allow cruise ship traffic, there should be contingency plans in place and money to pay for emergency equipment in communities" (George 2010). It is clear in the *Nunatsiaq News* coverage of the episode that the people of Kugluktuk had to scramble to prepare for the arrival of the rescued passengers.

The story of the *Clipper Adventurer* served as a warning to those involved in the voyage of the *Crystal Serenity*, highlighting the challenges posed by increased shipping and tourism in the Canadian Arctic as the ice pack shrinks and the waters of the Northwest Passage remain ice free longer each year. Canada claims sovereignty over the contiguous ice and waterways stretching to the North Pole based on human activity, including hunting on the ice, and navigation. In recent years, the Canadian government has also established national parks in remote places such as Baffin and Ellesmere islands (Timothy 297). While the authority of the Canadian state is ever present in the lives of people in the North, Canada is constantly preoccupied with national sovereignty in the form of control over Arctic waterways. In making claims for its ownership of Arctic waters, Canada has come late to the trend in Arctic tourism that has been developed in many parts of the Arctic. According to Dallen J. Timothy, the decline of the fishery and mining on Svalbard and the constant territorial pressure exerted by Russia, prompted the Norwegian government to promote tourism in order "to reaffirm its legal rights to Svalbard" (Timothy 295). The Arctic cruise industry developed in Canada in the 1980s, much later than in Scandinavia which saw its first Arctic cruise in 1845, and it developed in "an *ad hoc* character, only overcome by community enthusiasm and support from sub-sectors of the industry such as the cruise ship industry" (Viken et al. 253). In Nunavut, for example, "expedition cruises" that feature short visits in smaller vessels contributed an estimated $2.124 million to the economy in 2006 when three voyages were made through the Northwest Passage, and in 2007, approximately 2113 cruise passengers visited (Maher 121). In 2016, the *Crystal Serenity* made its voyage, fortunately without an incident.

The Arctic, long perceived primarily as a wilderness landscape visited by a daring few, has become a major tourist destination, even justifying the description of "mass" tourism used by tourism scholars.[2] There are many reasons for the surge in Arctic tourism including changes in global tourism (Grenier 4) such as the "general trend towards more specialized forms of leisure and growing consumer demand for new experiences" (Viken and Granås 2014).[3] The effects of this expanded travel in the Arctic have been mixed. Ultimately, climate change has not only made the Arctic more easily accessible, it has also changed how it is imagined as "the recurrent use of polar images to translate the abstract concept of climate change has not only brought a wide media attention to these forbidden regions but also allowed the drawing of a better understanding of these distant worlds" (Grenier 10). Ironically, Arctic cruises and other tourist activity leaves an enormous carbon footprint—creating what Byers calls a carbon "feedback loop"—and despite the national interests it seems to serve, cruise tourism is "literally a flow of technology, people, information, images and money, resulting in experiences produced on or nearby water almost anywhere on the globe" (Sletvold 175). As an industry, cruise tourism exemplifies "globalization," its liquidity and its consequences.

Even though Canadian tourism in the Arctic is not as well developed as the Norwegian industry, Canada is encouraging tourism as a way to bolster its claim to the Northwest Passage (Timothy 297). Norway joined other Arctic nations in exploring Arctic tourism as a means of asserting its sovereignty in the 1980s: "Before the 1960s, all of the legal requirements for sovereignty were being justified by claimant states through a history of resource extraction (eg mining and fishing), exploration, and scientific research. However, in the post-industrial world, tourism has also come to the fore in geopolitics, essentially replacing extractive economies with service economies as a foundation for territorial claims" (Viken 1995, 298). As Adriana Craciun and others note, Canada's assertion of sovereignty over the contiguous ice and waterways stretching to the North Pole has not been recognized by any other nation. In recent years, cruise tourism has begun to flourish, thus joining the two justifications for Canada's ownership of Arctic regions, human activity and navigation. While Arctic sovereignty preoccupies Canadian discussions of international relations as witnessed by the constant flow of books and articles on the subject, the role of tourism in the national story has yet to be analyzed fully. With each visit to one of the National Parks and every Canadian ship passing through the Northwest Passage, the argument for ownership based on use gains strength. The Arctic region off northern Canada was a cruise destination before the voyage of the *Clipper Adventurer* in 2010 (Stewart et al. 74, 77), yet as cruise tourism is expanding, traveling northward remains a relatively rare activity for the majority of Canadians who live in cities near the southern border. In contrast, traveling south for health care, education, or employment is common for people living in Arctic and sub-Arctic regions. Both types of travel perform the legacy of colonialism: one, the privilege of the south to go North, to seek out new experiences and to return; the other, the various forms of control that the state retains over those living in the North. What these developments mean for the Indigenous people of the Arctic remains a challenging question.

Literary Tourists and Canadian Identity

In *Northern Experience and the Myths of Canadian Culture*, I took a particular, perhaps one-dimensional, view of the tourists, including writers, who travel north, the writers who made brief journeys into the Arctic and returned south to write about it. Of course, these Canadians do not see themselves as tourists, but as visitors on "sojourn," the term Joan Sangster fruitfully applies to the archives many of them left behind (Sangster 2016 *passim*). The view of Arctic and Northern travel by literary tourists articulated what still seems to me to be a strong sense of entitlement. Tourism relies on a sense of otherness: the destination, the people, the culture a tourist visits bears some form of difference from the tourist's own experience. Perhaps this is why so many Canadians have traveled north without thinking of themselves as tourists: because the North, including the Arctic, is not considered "other" to Canada; it is considered the authentic experience of Canada.[4] For the Canadian northern enthusiast, the allure of the North is in part the role it plays in this imagined national identity. Therefore, the Canadian citizen, particularly the nationalist who believes that Canada's distinct identity rests on belonging to "the true North, strong and free," will imagine traveling in the Arctic, not as tourism, but as visiting another part, albeit a remote and unknown part, of their own land. Through the twentieth century, Canadians living in the south who went north went either to work or to study, often to study the place and its people. They were anthropologists, geologists, glaciologists, biologists— eventually literary scholars and writers went too. There was a time, before cruise ships sailed the Northwest Passage, when a trip to the Arctic was enough to justify the claim of knowledge and experience of the North.

"Literary tourists" was the term coined by Aron Senkpiel that I adopted to refer to the writers who produced a particular form of Canadian discourse that was very prominent at the time (Hulan 2002, 170). I took it further by calling northern enthusiasts the "cult of nordicity," a term I would regret, but only a little (Hulan 2002, 181). After all, there was a tendency among those who loved "the North" to exhort their fellow Canadians to "grasp" their northern identity as Rudy Wiebe does in *Playing Dead* without recognizing the colonial and colonizing gesture (see Hulan 1996). What made these travelers special, and what made them write about their travels, was how unusual it was. The typical Canadian traveler goes south not north. To go to the North required the courage to not follow the crowd; it required planning and, above all, money. Upon return, these writers claimed that visiting the North gave them invaluable firsthand knowledge, a special insight into the place that gives all of Canada a distinct identity, and thus, filled with patriotic spirit, they encouraged others to embrace true north. In contrast, seeing the writers as literary tourists highlighted the fact that no matter what the intention or outcome of travel northward, visitors are essentially consumers. The tourist is primarily a creature of appetite, consuming experiences as well as resources; the literary tourist takes these experiences and returns home to write about them.

In 2010, Kathleen Winter was a guest writer onboard the *Clipper Adventurer* when the ship ran aground in the Northwest Passage. In 2014, she published her story in the personal memoir *Boundless*. As the title suggests, the memoir charts a personal quest for an Arctic that is not contained within national borders or boundaries. Yet, to write a personal experience of travel is never to write only about the self. Wherever a traveler goes, there are other people. The North is inhabited by people, mostly Indigenous people. How to write about the experience of visiting a place raises the ethical problem of speaking about and for others. Conscious of her own history as a fairly recent settler, a history she weaves through the narrative, Winter asks, "My goal was not to colonize or subjugate the body of land. Or was it?" (149). Winter questions her own motivation, self-consciously reflecting on her own consumption, including her need to know the name of people she has met, to take home the dolls, carvings, and stories they have made. Often, she sees her curiosity as impertinence, a lack of respect, but she leaves the question of why she feels compelled to act as she does open until the end. Although self-consciousness is nothing new in Arctic travel writing, inflecting the voices of writers Rudy Wiebe and others, Winter's search for a sense of 'belonging' outside borders shows a trend in Canadian literary culture away from the true "Canadian" identity explored in the 1980s and 1990s toward circumpolar and global perspectives.

Boundless begins with a phone call from Noah Richler who asks if she knows that Russian icebreakers sometimes traverse the Northwest Passage: "They like to have a writer on board," he continues, and since he is unavailable, would she like to go in his place (5)? The invitation brings a flood of images. "I thought of Franklin's bones," she writes, "of the sails of British explorers in the colonial age, of a vast tundra only Inuit and the likes of Franklin and Amundsen and a few scientists had ever had the privilege of navigating … I thought of Queen Victoria and Jane Franklin, and of the longing and romance with which my father had decided to immigrate to Canada. I thought of all the books I'd read on polar exploration, on white men's and white women's attempts to travel the Canadian Far North" (5–6). She remembers learning to play the melody of "Lady Franklin's Lament." Finally, she agrees because "when a man called Noah suggests you get on a ship, hadn't you better jump on board" (7)?

Even though the reported conversation between Kathleen Winter and Noah Richler does not mention Adventure Canada, or the fact that the voyage is a cruise, Winter's amusing allusion identifies the voyage unwittingly and accurately with "extinction tourism," the kind of travel that seeks out species and places before they vanish forever. The Arctic tourist industry revolves around endangered animals, such as polar bears, and "themed destinations" based on Indigenous culture and community life (Viken and Granås, 6). These tours enlist a variety of experts to animate the destination. Billed as an "Arctic educational voyage" (17), the *Clipper Adventurer* hosted a "who's who" of Arctic experts including filmmaker and Inuit art dealer John Houston, geologist Marc St-Onge, writer Ken McGoogan, and musician Nathan Rogers, the son of balladeer Stan Rogers. Earlier in 2009, McGoogan had blogged about the voyage on the Adventure Canada website, describing the history of the area and the course the ship would take. Two Inuit women, Aaju Peter and Bernadette Dean (who also appears in John Walker's film *Passage*), were employed to provide cultural activities on board and to serve as gunbearers when out on the land. The cruise stopped in communities in Greenland, Nunavut, and the Northwest Territories. At Gjoa Haven, an art show and dance were organized; indeed, the community of Kugluktuk was preparing for such an excursion when the ship ran aground. Hired as the "resident writer," Winter was not given any teaching duties, unlike these other members of the "resource staff," the famous Arctic experts who gather passengers around to share their knowledge. The voyage is never described as a "cruise" in *Boundless*, and Adventure Canada is not mentioned by name. The ship is always referred to as "our ship," and only in a *Macleans* article does she call it a "cruise ship." In the media coverage of the ship's grounding, no mention is made of the famous passengers, and it is hard to say whether reporters counted them among the passengers or employees of the vessel, though the use of the term "crew" would suggest the former. Once on land in Kugluktuk, the passengers were not differentiated or identified in the media reports, and it does not seem that other experts have written about the incident.

As the initial invitation suggests, Winter and Richler are friends whose lives intersect in various ways: Richler's wife Sarah MacLachlan works at Anansi, the publisher of *Boundless* and Winter's brother Michael provided illustrations for Richler's *This Is My Country, What's Yours?* Yet, their public statements about the Arctic diverge significantly. While Richler maintains views steeped in Canadian nationalism, Winter seems to want to transcend it. Richler's position seems to come from the idea of "Canada as North" behind Canada's continued assertion of national sovereignty over the Northwest Passage. In *This Is My Country, What's Yours?*, Richler charts his search for a sense of national identity: "Being Canadian demands a constant effort of the imagination, a working definition of the country must be conjured out of the ether on consecutive mornings" (455). The North figures prominently in the search which takes him to such places as Iqaluit, Igloolik, and Inuvik. He does not mention Mordecai Richler's classic Arctic novel, *Solomon Gursky was Here*, perhaps seeking to stake out a claim to Northern territory separate from his father's. By the end of the journey, he concludes, "Canada is a fiction, ours to invent" (459). This conclusion frees him from responsibility to justify the claims being made and allows him to maintain skepticism revealed in an article he published concerning the Franklin expedition:

> The moral of the Canadian telling is that had Franklin paid attention to Inuit inhabiting the region—had he been a good and conscientious fella of the sort that Canadians imagine themselves to be today—then he would have heeded the Inuit and their *qaujimajatuqangit*, or "traditional knowledge," and survived. The story of the loss of Franklin and his ships was a Canadian foundation myth, a vessel for what was felt to be Canadians' better nature, regardless of the plight of indigenous peoples here (that's

how a good myth works). The story, as it was told, was a harbinger of the country that would not come into being for another couple of decades. (Richler 2014)

As a journalist, Richler often crafts a tone of world-weariness, though the rhetoric can get him into trouble.[5] The choice of words in this article risks sounding dismissive; for instance, the skepticism conveyed by placing quotation marks around Traditional Knowledge seems not only to question the Inuit accounts but perhaps their entire worldview, if not Indigenous worldviews generally. He also seems to disdain those political allies and sympathizers who he says "have used the occasion to *clamour* on behalf of the Inuit, their *qaujimajatuqangit* more accurate about the location of the ships than 170 years of science were" (emphasis added). Richler also leaves out Inuit history and dismisses Franklin's cultural legacy: "this is what remains of the Canadian Franklin story, the last possibility of the indigenously crafted sensibility of the once colonized nation having been superior" (Richler 2014). Doubts concerning Inuit accounts of the Franklin expedition are nothing new of course. The attitude originated in England in the Victorian period, and many Canadians remain ignorant of Inuit contributions to historical knowledge. Nevertheless, the "moral" of the Franklin story is no mere figment of the politically correct imagination of Canada's intellectuals.

As Adriana Craciun reminded readers of the *Ottawa Citizen* after the discovery of the HMS Erebus: "The Franklin expedition did not die for Canadian sovereignty," and the involvement of Shell as a partner in the expedition to locate the ships illustrates how scientific exploration has been entangled in a web of politics, involving industrial, commercial, and military interests since the early modern period (Craciun 2014; see also Craciun 2016). In response to her article, Richler dismissed Craciun by characterizing her as "an academic trained in the UK and teaching in California" and disparaged her proposal for UNESCO designation calling it "patronising not only towards Canadian science and the country's pioneering system of national parks, but towards the country itself, if in a manner to which Canadians are well accustomed" (Richler 2014). For Richler, Craciun's affiliations with Britain and the United States (not to mention her American citizenship) render her suspect, an interloper meddling in Canadian affairs. As a Canadian, he chastises her attitude "towards the country itself," and says nothing of her academic qualifications or years of studying the literature and culture of the Victorian period culminating in the study *Writing Arctic Disaster* (Craciun 2016), preferring the authorization of national identity over knowledge and expertise. It is ironic that this argument arises in an article exhorting another country, Britain, to heed lessons learned from Canada, and it betrays the nationalism that lingers despite the shift toward more cooperative, global, and circumpolar perspectives evident in Canadian writing about the Arctic.

In *Boundless*, Winter seems to reject the Canadian nationalist stance in favor of "the possibility of 'belonging' outside borders" (56). At first, she sets out on a quest to find the feeling she had as a child when "something tantalizing and immaterial made itself known to me, glinting from beyond the ordinary world with which grownups seemed so contented" (127). The search for a crack in the material world revealing the imminence beyond it does not end but is transformed into the feeling that the land is speaking, and her goal becomes learning to listen and to understand what she comes to call "true sovereignty," an authenticity that comes from the land itself. This search for knowledge only briefly evokes Arctic literary history when she refers to "all the books" she has read and credits two titles, *This Cold Heaven* by Gretel Ehrlich and *The Last Imaginary Place* by Robert McGhee. As in other Arctic narratives, the narrator seeks to shed the baggage carried into the Arctic, including the past, becoming "boundless," freed from all constraints. In Winter's case, this means a personal quest for healing and vision. We learn that her first husband died young of a terrible illness, leaving her and their young daughter. We hear her family history, her parents' decision to emigrate from England to Newfoundland when she was eight years old; their life in the relatively secluded place her father equated with freedom; her life in various outports; her first husband's death; her move from Newfoundland to Montreal with a new husband and second daughter. Still making sense of these events, she discovers that she needs to listen to others and to the land. What she eventually learns is summarized in the article she wrote for *Macleans* magazine:

> Canada's Arctic is a place of sapphires and energy, power and true sovereignty, deeply important to our collective future. How we listen to it and to its original inhabitants may determine our salvation or our doom. The uncharted rock on which our ship ran aground is just one fragment of an entire world we have yet to perceive. (Winter 2010)

This sense of the Arctic as a different world, an absolute otherness defying understanding is also a literary trope, but what is different is the context in which she places the experience and the direction she takes after the voyage. In her review of *Boundless*, Joanna Kavenna wishes for the concrete data that Winter avoids, such as the "maligned 'tonnage of ice-bergs,' as well as the inhabitants, flora, and fauna," but notes that "Winter's concern is not with those things but with 'the emotional reality' of her journey, which 'touched another realm'" (24). This is a perceptive reading, for although the narrator observes the people around her, she remains apart. She relates that she "gravitated toward" the two Inuit women on board, Aaju Peters and Bernadette Dean: "I wanted to hear more of what Aaju Peter had to say about this [Inuit history], as a Greenlandic woman who had adopted a Canadian Inuit life. I wanted to ask her about the possibility of 'belonging' outside borders" (56). While she avoids the history buffs and geologists generally, citing her own obtuseness particularly about geology, each individual does appear in the story as she relates to the people and the place intuitively, having made a conscious choice to search for "something tantalizing and immaterial" (127).

In *Boundless*, the narrative of exploration and discovery is constantly undermined by the narrator's avoidance of the other experts. It is not from lack of respect or friendship for the individuals: Winter's accounts of her interaction with Marc St-Onge and Ken McGoogan are warm and genial, and her questions to them reveal rare but genuine interest. McGoogan is the person she seeks out in Gjoa Haven immediately after a local man named Wally Porter confides in her that the community will soon unearth a site believed to contain Sir John Franklin's logbook, and she writes approvingly of both McGoogan's treatment of her revelations and his reporting of the story. In another moment, she describes her personal focus of study as the muskoxen, the land, the bear spotted on Beechey Island. Onboard, she drifts away from most of the organized activities, the one exception being, listening to Stan Rogers' "Northwest Passage" sung by his son Nathan, who she befriends. Throughout the book, she maintains her distance from the Arctic explorer image cultivated by the cruise industry though she comes to appreciate the knowledge she can gain from some of them, especially after the grounding, from geologist Marc St-Onge in particular. It is clearly not a personal aversion or a lack of respect for their knowledge that makes her avoid the experts; rather a search for something else leads her away. Early in the memoir, she expresses her desire for different knowledge than the experts have to offer:

> On the land I relied on my solitude, my walking and observations, but I also gravitated toward Bernadette Dean and Aaju Peter because I wanted their Inuit and Greenlandic perspectives—I wanted to hear what women of this land had to say, and was less interested in the old European male Arctic explorer stories to which the history buffs thrilled. (33)

Winter is clearly uncomfortable with the explorer figure evoked by the Arctic tourist industry.

As she packed for the expedition with "a list in hand from the expedition leaders," Winter describes how she came across photos of the other "resource staff" hired for the trip: "I noticed they were nearly all men, and most had explorer-type beards" (9). Rather than comment at that point in her narrative, Winter gently satirizes the whole affair: "I happened to have a beard I'd crocheted out of brown wool on a train trip with my mother— it was a bit more Rasputin than Explorer, but it possessed loops that fit nicely around my ears, so I packed it as well" (9). When would a woman ever wear such a thing, the reader may wonder, but once the importance of needlework becomes apparent in the narrative, the crocheted beard takes on special significance, literally knitting together her skepticism toward scientific and historical knowledge represented in the figure of the "buff" and her desire for the knowledge that comes from experience of the material world represented by the land, the women, and their work. When she finds the occasion to wear her beard, it is a playful gesture, a satire on the imagery of rugged masculinity associated with exploration. It comes after the visit to Gjoa Haven. In the evening after she brings the story of Franklin's logbook to McGoogan, she finally dons her explorer's beard: "I was still a tiny bit proud of myself for being the conduit through which the story had decided to flow, and at dinner I felt it was time, finally, to wear my crocheted explorer's beard, attached snugly and effectively to the face by way of discreet loops around one's ears" (208), and making the joke complete, she concludes: "I'm sure the beard did a lot to cement my reputation, among the ship's scholars, as a true finder of important Arctic secrets" (208).

Winter's send-up of Arctic exploration underlines her earlier discomfort with the commodified knowledge the cruise is selling. At least part of the narrator's reticence toward expertise seems aimed at the myth of discovery and the stereotyped masculine Explorer. Understandably, she is drawn to the women, but especially those who have particular knowledge. After attending a workshop led by artist Sheena Fraser McGoogan, Winter remains in the library: "I'd brought my crochet needle and a skein of wool my friend Marilee had given me, spun and hand-dyed by Shawn O'Hagan in Newfoundland" (54). The activity connects her to women artists and friends while also bridging the social space of the workshop and the loneliness of the Arctic that she is seeking. Like stories, handmade garments are works of art and artisanship, vision, and practicality. This is made clear in the scene where she admires the "undergarment" that her fellow passenger and friend Elisabeth carries with her: "Of web-fine wool, it possessed long sleeves ending in demure shirred edges, a scalloped neckline, and a pattern of wild roses. It looked like something Lady Franklin might have worn" (93). Winter admires the delicacy of this garment so much that she also persuades Elisabeth to let her borrow it to sketch, the better to know the fine weave of its wool. Arts and crafts are a constant theme throughout the book that joins the work of the writer on board the expedition to historic forms of labor such as the sewing, knitting, and leatherwork performed by generations of Indigenous women. Sitting on deck knitting, working strands of the muskox hair she has collected into her work, she presents an image of domesticity in the midst of adventure in a way that recalls the multiple forms of labor that supported British and Canadian Arctic travel. Onboard, Bernadette tells stories of her grandmother Shoofly who fell in love with the American whaling captain George Comer and whose bead-work *tuilli* and other garments are still held at the Museum of Natural History in New York along with the meteorites that Robert Peary took from Greenland. Shoofly's name was Niviatsianaq, and she is well-known in Arctic history, if not for these particular garments, for the often repro-duced photo of her sitting at her sewing machine, reputed to be the first sewing machine used by Inuit women. Winter is so intrigued that after the voyage that she travels to the American Museum of Natural History to see her garments even though the curator tells her they can be viewed online. Shoofly's needlework means more to her as a relic of Arctic exploration than the bones and buttons left by Franklin and crew. Through these gar-ments, another history is perceived, a history of colonial encounter through the labor provided by generations of women. While Winter is avid for the perspectives of the Inuit women, the narrative tends to resist cultural appropriation because what she learns from the women is always repre-sented as theirs. She is the pupil and observer to their teachings. If *Boundless* occupies discursive space that might have been taken up by Inuit speakers, it also creates space for the work of decolonization that should be the pri-ority of all writers. As Margery Fee argues, the alternative to speaking about Indigenous knowledge can be to "support continuing colonization" (38). From her original longing for a "glimpse" of the immanent world that motivates her adventure to the experience of being in the Arctic, Winter develops a new desire for an understanding of the land that can only be learned from the people who know it. As she becomes aware, the carefully woven text gradually unravels in the final chapters. It is the beginning of Winter's attempt to "unsettle the settler within," as suggested in Paulette Regan's evocative title, and it leads to the disintegration of the narrative into short episodic glimpses of a learning process that has only begun. As a literary tourist, she continues her journey as she tries to learn more about the people and places she has been, and knowledge becomes the main theme of the book's open-ended conclusion.

By dramatizing her personal discovery of her position within settler colonialism, Winter carries on the literary work of those writers who manage to define and also challenge the position of the literary tourist. The prime example of this literature, and one worth pausing to reconsider in this context is *North of Summer*, Al Purdy's collection of poems inspired by his trip to Baffin Island in 1965. A classic of the Canlit period, the collection has long been embraced and critiqued as an encounter with the remote and alien world of the Arctic. As Lorraine York observes, the collection turned Purdy into "*the* Northern poet."[6] Indeed, the trip to the Arctic was so important to Purdy that he took a phrase from Stan Rogers' song "The Northwest Passage" for the title of his autobiography: *Reaching for the Beaufort Sea*. In it, he describes his trip to Baffin Island in 1965 as a new beginning for him as a poet: "No other poet I knew of had ever gone to the Arctic (except Robert Service, and he didn't count), it was virgin territory for me, untouched except for the mundane prose of explorers and scientists" (Purdy 190). Here was a chance to go where others had not been both literally and figuratively (though the book was dedicated to Frank Scott, a poet who had visited the Arctic.) Nevertheless, it would allow him to stake a claim to his own poetic territory, as York shows, creating the "North as icon" in his poetry. Purdy's stake in the "true North" was further solidified by his concern for authenticity. In a letter to George Woodcock, he complained when his publisher chose A. Y. Jackson's paintings to illustrate *North of Summer* instead of Inuit prints. The North depicted in these paintings was not the one that had captured his poetic imagination, nor did the imagery match his admiration for the people he had met. The Group of Seven's lonely, mystical northern wilderness was by this time a staple of Canadian nationalism, and Purdy knew that the publisher was using them to sell the book (Van Rys 3).

Purdy had tried to unlearn these ways of seeing the North in his encounter with the Arctic, and his search for knowledge shapes many of the poems. In "Trees at the Arctic Circle," for example, he dramatizes the arrogance of the outsider's viewpoint and records his initiation into the Arctic world. The speaker, an outsider, must acknowledge assumptions about the Arctic trees in order to truly experience. At first, the speaker contrasts the tiny Arctic willows to their southern relatives, the towering Douglas firs that seem so proud in comparison, but then, he realizes that the Arctic trees "make life from death," and once he learns how tough and resilient the Arctic trees they are, he makes his own discovery that he has been "stupid in a poem" (Purdy 1967, 30). This revelation reveals the vulnerability that Ian MacLaren observes in the collection as a whole and remains as confirmation of the knowledge the speaker seeks to acquire.[7] Having acquired knowledge through this experience, the speaker can be imagined as moving from outside to inside Arctic reality.

While Purdy's speaker confronts a lack of knowledge of the land and its people in this poem, drawing on the poet's experiences traveling Baffin Island, the Romantic figure of Kudluk in his poem "Lament for the Dorsets" owes more to literary traditions of imagining Indigenous peoples than to firsthand experience. As they are seen "squatting" at their lamps and scratching their heads with their "hairy thumbs," the primitive Dorset people seem ready to disappear into the past, but for the longevity of the art they leave behind. The speaker wonders "Did they ever realize at all / what was happening to them?" (Purdy 1968, 55). Unable to adapt to the arrival of the Thule or to the change in climate, the fictional Dorset people seem fated to disappear. Indeed, the poem seems to echo the Romantic image of the mother slowly blanketed in snow in Duncan Campbell Scott's "The Forsaken," a poem that projects the "dying Indian" trope. Similarly, "Lament for the Dorsets" mourns the disappearance of those known to archaeologists as the Dorsets, believed to have become "extinct," possibly wiped out by the arrival of the Thule people in approximately 1000 AD, though their fate remains the subject of debate. Purdy elegizes while firmly situating the early Arctic people in a past that has been surpassed by the modern world.

In this poem, the speaker tries to imagine ancient Arctic people as he contemplates artifacts on the Arctic land. As Sandra Djwa observes, "Purdy imagines an evolutionary structure gone wrong in which the Dorsets and their camp are extinguished through competition and climate change" (Djwa 60). The poem probes the problem of how to imagine people whose lives are so remote in time by conjuring the image of one individual, "the last Dorset," carving the tiny ivory swans later scattered on the land. In his mind's eye, the speaker watches the carver reach the moment of creation when a thought "turns to ivory." The carving completed, Kudluk succumbs to the elements; his tent is blown away, and the poem concludes with the knowledge that centuries later, the thought is "still warm" (1968, 55). "For Purdy," Djwa observes, "it is the attraction of the work of art as a living thing" that continues to resonate across time (Djwa 60). This "ivory thought" captured the critical imagination, furnishing several titles on Purdy's work, including the proceedings of a Reappraisals conference at the University of Ottawa in 2006. The focus on the ancient artifact also resonated with the obsession with "digging up of buried things" that Margaret Atwood observes in Canadian culture (1997, 19). The poem imagines cultural survival assured through art, especially the artist's ability to reconcile the life of the imagination and the real, to make experience meaningful. But, for the modern poet, the imaginary Kudluk and his imaginary granddaughter must die for this transmission to be realized. The "dying Indian" trope shadows Purdy's search for new knowledge, foreclosing the chance to learn from the ancient Inuit, and the image of the Dorsets' profound ignorance of the world, their helplessness and hopelessness in the face of change, conflicts with what we now know of the passing on of Indigenous knowledge, and the adaptability of Arctic peoples. Indigenous writers generally dispute the division of tradition and Modernity that has been used to marginalize their culture, insisting that Modernity and tradition have always coexisted.

The distinction between insider and outsider knowledge in Purdy's experience of initiation into the Arctic world shaped the literary reception of Arctic narrative for decades. In large part, this distinction was based on the assumed dichotomy of tradition and Modernity that cast Indigenous societies generally, and Inuit societies most blatantly, as pre-modern (see Hulan 2016). Purdy was not immune to this view of the Inuit. Today, the dichotomy is being rapidly broken down by the experience of climate change. The global significance of the changes to local environments in the Arctic is dissolving conceptual borders as insiders and outsiders share a common interest in what happens to the earth. At the same time, the global community can learn from those whose lives have been affected most by climate change, especially those with the deep knowledge of the land and water that comes from tradition. To understand this relationship, readers can also learn from the ongoing engagement with local narratives communicated in studies like Julie Cruikshank's *Do Glaciers Listen?* After years of talking to storytellers and listening to the stories of glaciers in the Saint Elias Mountains, Cruikshank concludes:

Narrative recollections and memories about history, tradition, and life experience represent distinct and powerful bodies of knowledge that have to be appreciated in their totality, rather than fragmented into data, if we are to learn anything from them. (259)

She calls these "entangled narratives" because the traditional stories are constantly evolving and changing, like the glaciers themselves, in response to the global discussion and generation of narratives concerning climate change. Michael Bravo draws similar conclusions by tracing the changes in northern scientific methods arising from the contest for "the authoritative spaces" and the arguments Indigenous societies have made for the equivalence between Indigenous knowledge and the knowledge generated in the natural sciences (Bravo 2000, 470; see also Berger 1988). As scientists transform their practices in light of these challenges, Bravo concludes that "the practices of science and Indigenous knowledge are thoroughly entangled and mutually constitutive rather than dichotomous" (Bravo 2000, 474). Emilie Cameron's *Far Off Metal River*, which rejects the practice of looking to Indigenous oral traditions for the counter-narrative to colonial discourse, demonstrates further how situated knowledges can be generated by creating a dialog between settler and Indigenous stories. Therefore, the "we" in Cruikshank's sentence above reflects the same "we" Cameron and others are writing for not the "we" of the settler scholar or the academy, but of people everywhere.

FROM TOURIST TO WITNESS

While documenting the cruise through the Northwest Passage, Kathleen Winter reflects on her life as an immigrant to Canada as she embarks on a new physical and emotional journey into contemporary Indigenous issues when she travels to New York to see the Greenland meteorites and the garments made by Niviatsianaq that Comer purchased for the museum in 1906 (Eber 115). Sitting by Anhighito (the Tent), the Woman, and the Dog, as the Greenlanders called the meteorites they had known for generations, she reflects on how they "speak using not words but substance" (248). One wishes that Winter had more to say about the failure to repatriate Niviatsianaq's beadwork, or the Cape York meteorites; more about the way tourism serves and does a disservice to the people of Kugluktuk and the other communities they visit. As a narrator, she is often more accepting than probing when it comes to these connections. Yet, she does not leave the reader in the Museum. Instead, she makes one more journey, a pilgrimage to Ottawa where Teresa Spence, then chief of Attawapiskat was on a hunger strike to protest Prime Minister Harper and Governor-General Johnston's refusal to meet with her to discuss the housing and education crises on her reserve. The brief passage describing Winter's visit maps a gradual acquisition of knowledge, and this discovery shorn of heroic language transforms the tourist into a witness. Winter knows that she lacks the lived experience of an Indigenous person and that she can never be as reliable a witness to Indigenous issues as those who live them. All she can do is to write her settler perspective, exposing her own lack of knowledge and describing her attempts to learn and to become an ally. At first, she cannot find Victoria Island, tucked behind the Parliament buildings, and she has no idea what she will do when she gets there. Invited to make an offering of tobacco, she admits that she does not know what that means, that she has not brought any. But, she has brought some Labrador tea to give to the Chief, hoping that the fragrant leaves hold the healing beauty of the land where they were picked, and she sits vigil with the others beside the fire, cringing as an aggressive young man posing as a journalist rudely questions the man who is keeping watch over the gate. The man's patience soothes her, and his answers give her a touchstone when, asked repeatedly if the hunger strike is being held on sacred land, the man replies: "ALL LAND IS SACRED" (259).

In 2016, media attention turned again to the community of Attawapiskat when the number of teens attempting or planning to commit suicide spiked. Even though the emergency response of the provincial and federal governments was swift, and some community members even spoke of a sense of hope, there were no new or concrete solutions. Life in Attawapiskat and most of the North remains testimony to the legacy of colonialism. The sacredness of the land that Winter learns may help readers grasp why calling on the people to move, as former Prime Minister Chrétien suggested at the time, is really no solution. As more tourists cruise Arctic waters every summer, looking for endangered wildlife and watching the dancing and drumming in picturesque Inuit communities, what do they learn about the living conditions of people in the Arctic and northern regions of Canada? With their cameras and phones poised to capture the vanishing Arctic, what do these visitors cruising the Northwest Passage discover? Winter's memoir leaves such questions unanswered yet offers one woman's attempt to learn from and to support Indigenous people and knowledge, one literary voice that is not preoccupied only with the adventures that have led to the "great derangement" of our time.

NOTES

1. The *Clipper Adventurer* was also used in 2007 by the Students on Ice program that James Raffan describes in *Circling the Midnight Sun* (Toronto: HarperCollins, 2014).

2. In *Tourism Destination Development*, Brynhild Granås and Arvid Viken, provide an analysis that is deeply influenced by the work of Doreen Massey on the relational nature of space and the concept of situated knowledge introduced by feminist ethnographers, exploring Arctic destinations through the material conditions of place, including often overlooked elements such as infrastructure and technology (London: Ashgate, 2014, 1–2), and contributing a conceptually rich discussion of destinations as sites for both consumption and production.

3. In *Polar Tourism: A Tool for Regional Development* Alain Grenier offers several more, ten in total, including: the end of the Cold War; the emancipation of Aboriginal peoples and subsequent devolution of authority to their communities; the identification of new regions like Nunavik; the shift from extraction to a service-based economy; the environmental movement; saturation of the market; the threat to mass tourist destinations posed by violence and terrorism; and, the media representation of a vanishing Arctic (Québec: Université Québec, 2011, 10–11).

4. This argument is developed in "'Everybody Likes the Inuit': Inuit Revision and Representations of the North" (*Introduction to Indigenous Literary Criticism in Canada*. Ed. Heather MacFarlane and Armand Garnet Ruffo. Peterborough: Broadview, 2016).

5. During the 2015 federal election in which Richler ran for the NDP in Toronto-St. Paul's losing to Liberal candidate Carolyn Bennett by over 23,000 votes (Bennett received 55% of the vote, the Conservative candidate Marnie MacDougall won 27%, and Richler came in third with 15%), he had to retract intemperate remarks on Facebook in which, among other wild accusations, he called then Prime Minister Harper a "psychopath."

6. In "Unsettling *North of Summer*," L. Camille van der Marel critiques Purdy's attempt to "indigenize" himself by taking on this persona and offers a postcolonial analysis of the collection (*Ariel* 44, no. 4 [2014]: 13–47).

7. See I.S. MacLaren's "Arctic Al: Purdy's Humanist Vision of the North" in *The Ivory Thought* (University of Ottawa Press, 2008, 128). In the same collection, Janice Fiamengo notes how these poems of "clumsy lament" have a "deliberately unfinished quality" and "foreground their verbal inadequacies before they risk a lyrical flight" (161).

REFERENCES

Arctic Cruise Company Sues Over Stranded Ship. 2011. *CBC*, 11 July. Accessed April 20, 2016. http://www.cbc.ca/news/canada/north/arctic-cruise-company-sues-over-stranded-ship-1.1083047

Arctic Rescue Fears as Massive Cruise Ship Prepares to Sail the Northwest Passage. 2016. *CBC*, 2 April. Accessed April 20, 2016. http://www.cbc.ca/news/canada/north/cruise-ships-safety-northwest-passage-1.3518712

Atwood, Margaret. 1997. *In Search of Alias Grace*. Ottawa: University of Ottawa Press.

Berger, Thomas R. 1988. *Northern Frontier, Northern Homeland*. Rev. ed., 2 vols. Toronto: Douglas and McIntyre.

Bravo, Michael. 2000. Cultural Geographies in Practice: The Rhetoric of Scientific Practice in Nunavut. *Ecumene* 7 (4): 468–474.

Byers, Michael. 2016. Arctic Cruises: Fun for Tourists, Bad for the Environment. *Globe and Mail*, 18 April. Accessed April 20, 2016. www.theglobeandmail.com/opinion/arctic-cruises-great-for-tourists-bad-for-the-environment/article29648307/

Cameron, Emilie. 2015. *Far Off Metal River: Inuit Lands, Settler Stories, and the Making of the Contemporary Arctic*. Vancouver: University British Columbia Press.

Coast Guard Seeks Damages for Arctic Cruise Ship Accident. 2012. *CBC*, 19 June. Accessed April 20, 2016. http://www.cbc.ca/news/canada/north/coast-guard-seeks-damages-for-arctic-cruise-ship-accident-1.1173325

Craciun, Adriana. 2014. Franklin's Sobering True Legacy. *Ottawa Citizen*, 10 September. Accessed October 5, 2015. www.ottawacitizen.com/news/national/adriana-craciun-franklins-sobering-true-legacy

———. 2016. *Writing Arctic Disaster: Authorship and Exploration*. Cambridge: Cambridge University Press.

Cruikshank, Julie. 2005. *Do Glaciers Listen? Local Knowledge, Colonial Encounters, and Social Imagination*. Vancouver: University of British Columbia Press.

Cruise Ship Exploring the Northwest Passage Runs Aground. 2010. *Globe and Mail*, 29 August. Accessed October 5, 2015. www.theglobeandmail.com/news/national/cruise-ship-exploring-northwest-passages-runs-aground/article1378559

Djwa, Sandra. 2008. Al Purdy: Ivory Thots and the Last Romantic. In *The Ivory Thought: Essays on Al Purdy*, ed. Gerald Lynch, Shoshannah Ganz, and Josephene M. Kealey, 51–62. Ottawa: University of Ottawa Press.

Fee, Margery. 2015. *Literary Land Claims: The "Indian Land Question" from Pontiac's War to Attawapiskat*. Waterloo, ON: Wilfrid Laurier University Press.

Fiamengo, Janice. 2008. Kind of Ludicrous or Kind of Beautiful I Guess: Al Purdy's Rhetoric of Failure. In *The Ivory Thought: Essays on Al Purdy*, ed.

Gerald Lynch, Shoshannah Ganz, and Josephene M. Kealey, 159–171. Ottawa: University of Ottawa Press.

George, Jane. 2010. Stranded Passengers Find Warmth in Kugluktuk. *Nunatsiaqonline*, 30 August. Accessed May 12, 2016. www.nunatsiaqonline.ca/stories/article/3008109_stranded_passengers_find_warmth_in_Kugluktuk/

———. 2012. TSB Report on Clipper Adventurer Grounding Reveals Broken Equipment, Questionable Decisions. *Nunatsiaqonline*, 27 April. Accessed April 20, 2016. www.nunatsiaqonline.ca/stories/article/65674tsb_report_on_the_clipper_adventurers_grounding_reveals_broken_equipme/

Granås, Brynhild. 2014. A Place for Whom? A Place for What? The Powers of Destinization. In *Tourism Destination Development: Turns and Tactics*, ed. Arvid Viken and Brynhild Granås, 79–92. Farnham, UK: Ashgate.

Grenier, Alain, and Dieter Müller. 2011. *Polar Tourism: A Tool for Regional Development*. Québec: Presses des Université Québec.

Hulan, Renée. 1996. Literary Field Notes: The Influence of Ethnography on Representations of the North. *Essays on Canadian Writing* 59: 147–163.

———. 2002. *Northern Experience and the Myths of Canadian Culture*. Montreal: McGill-Queens University Press.

———. 2016. Everybody Likes the Inuit': Inuit Revision and Representations of the North. In *Introduction to Indigenous Literary Criticism in Canada*, ed. Heather MacFarlane and Armand Garnet Ruffo, 201–220. Peterborough, ON: Broadview.

Kavenna, Joanna. 2015. Boundless: Tracing Land and Dream in a New Northwest Passage. *Times Literary Supplement*, 24 April, 24 [Revised].

MacLaren, I.S. 2008. Arctic Al: Purdy's Humanist Vision of the North. In *The Ivory Thought: Essays on Al Purdy*, ed. Gerald Lynch, Shoshannah Ganz, and Josephene M. Kealey, 119–136. Ottawa: University of Ottawa Press.

Maher, Patrick T. 2010. Cruise Tourist Experiences and Management Implications for Auyuittuq, Sirmilik and Quttinirpaaq National Parks, Nunavut, Canada. In *Tourism and Change in Polar Regions: Climate, Environments and Experience*, ed. C. Michael Hall and Jarkko Saarinen, 119–134. New York: Routledge.

Marine Investigation Report M10H0006. Accessed October 12, 2015. http://www.bst-tsb.gc.ca/eng/rapports-reports/marine/2010/m10h0006.asp/

Purdy, Al. 1967. Trees at the Arctic Circle. In *North of Summer: Poems from Baffin Island*, 29–30. Toronto: McClelland.

———. 1968. Lament for the Dorsets. In *Wild Grape Wine*, 54–55. Toronto: McClelland.

———. 1993. *Reaching for the Beaufort Sea: An Autobiography*. Madeira Park, BC: Harbour.

Raffan, James. 2014. *Circling the Midnight Sun: Culture and Change in the Invisible Arctic*. Toronto: HarperCollins.

Regan, Paulette. 2010. *Unsettling the Settler Within: Indian Residential Schools, Truth Telling, and Reconciliation in Canada*. Vancouver: University of British Columbia Press.

Richler, Noah. 2006. *This Is My Country, What's Yours? A Literary Atlas of Canada*. Toronto: McClelland.

———. 2014. What Canada—And John Franklin—Can Teach the UK About the Independence Game. *New Statesman*, 16 September. Accessed May 20, 2016. http://www.newstatesman.com/politics/2014/09/what-canada-and-john-franklin-can-teach-uk-about-independence-game

Sandiford, K. 2006. Cruise Control. *Up Here*, May/June, 38–43.

Sangster, Joan. 2016. *The Iconic North: Cultural Constructions of Aboriginal Life in Postwar Canada*. Vancouver: University of British Columbia Press.

Sletvold, Ola. 2014. Standardization and Power in Cruise Destination Development. In *Tourism Destination Development: Turns and Tactics*, ed. Arvid Viken and Brynhild Granås, 171–188. Farnham, UK: Ashgate.

Stewart, E.J., and J. Dawson. 2011. A Matter of Good Fortune? The Grounding of the *Clipper Adventurer* in the Northwest Passage, Arctic Canada. *Arctic* 64 (2): 263–267.

Timothy, Dallen J. 2014. Contested Place and the Legitimation of Sovereignty Claims through Tourism in Polar Regions. In *Tourism Destination Development: Turns and Tactics*, ed. Arvid Viken and Brynhild Granås, 288–300. Farnham, UK: Ashgate.

Van der Marel, and L. Camille. 2014. Unsettling *North of Summer*: Anxieties of Ownership in the Politics and Poetics of the Canadian North. *Ariel* 44 (4): 13–47.

Van Rys, John. 1990. Alfred in Baffin Land: Carnival Traces in Purdy's *North of Summer*. *Canadian Poetry* 26: 1–18.

Viken, Arvid. 1995. Tourism Experiences in the Arctic—The Svalbard Case. In *Polar Tourism: Tourism in the Arctic and Antarctic Regions*, ed. C.M. Hall and M.E. Johnston, 73–84. Chichester, UK: John Wiley & Sons.

Viken, Arvid, and Brynhild Granås. 2014. Dimensions of Tourism Destinations. In *Tourism Destination Development: Turns and Tactics*, ed. Arvid Viken and Brynhild Granås, 1–17. Farnham, UK: Ashgate.

Wiebe, Rudy. 2003. *Playing Dead: A Contemplation Concerning the Arctic*. Edmonton: NeWest.

Winter, Kathleen. 2010. An Arctic Accident. *Macleans*, 16 September. Accessed October 10, 2015. www.macleans.ca/society/life/an-arctic-accident/

———. 2014. *Boundless: Tracing Land and Dream in a New Northwest Passage*. Toronto: Anansi.

York, Lorraine. 1993. The Ivory Thought: The North as Poetic Icon in Al Purdy and Patrick Lane. *Essays on Canadian Writing* 49: 45–56.

CHAPTER 3

Negotiating Sovereignty: Sheila Watt-Cloutier's *The Right to Be Cold*

Abstract As an Inuit woman who maintains a worldview rooted in Inuit culture and Indigenous knowledge, Sheila Watt-Cloutier brings her knowledge of Inuit traditions and lived experience into the international world of diplomacy and public policy. *The Right to Be Cold* is her memoir of the struggle to make climate change a priority in the international community. In it, she argues that the protection of the ways of life depending on the Arctic environment, particularly the cold, is a human right that transcends national boundaries. Hulan argues that Watt-Cloutier's writing not only shows how Indigenous leaders from the Canadian Arctic are exerting their own sovereignty and forging alliances with others from the circumpolar world and beyond to pressure governments to take action to protect Indigenous communities and traditions threatened by climate change, but also teaches readers about Indigenous knowledge.

Keywords Climate change • Inuit autobiography • Indigenous knowledge • Leadership • *Sila*

The "great derangement" described by Amitav Ghosh was on public display in the 2017 CBC *Canada Reads* contest. The game was framed around the question: "What is the one book Canadians need now?" Five celebrities would choose a book, and over five days, the participants would debate the merits of their choices, voting one book off at the end of each

© The Author(s) 2018
R. Hulan, *Climate Change and Writing the Canadian Arctic,*
https://doi.org/10.1007/978-3-319-69329-3_3

show. Among the books proposed by the celebrity contestants was an unlikely but potent choice: *The Right to Be Cold* by Sheila Watt-Cloutier. *The Right to Be Cold* is a memoir of the 20 years Watt-Cloutier spent as an elected leader and advocate working to move climate change to the forefront of the international agenda. It documents the role Indigenous peoples are playing in bringing attention to climate change and negotiating measures to deal with it at the international level. In these negotiations, Indigenous leaders from the Canadian Arctic are forging alliances with others from the circumpolar world as well as with environmental activists and Small Islanders from around the globe to pressure governments to take action on global warming and thereby to protect Indigenous communities and traditions threatened by climate change. Watt-Cloutier argues that the protection of the ways of life depending on the Arctic environment, represented by the cold, is a human right that transcends national boundaries.

As a work of non-fiction, *The Right to Be Cold*, chosen and defended by musician Chantal Kreviazuk, stood out from the novels chosen by the other celebrity panelists Measha Brueggergosman (*Company Town* by Madeline Ashby), Humble the Poet (*Fifteen Dogs* by André Alexis), Candy Palmater (*The Break* by Katherena Vermette), and Jody Mitic (*Nostalgia* by M. G. Vassanji). From the very first debate, it was clear that *The Right to Be Cold* would remain an outlier on *Canada Reads*. Although echoed in comments made by the other panelists, Jody Mitic was the book's toughest critic: "When asked whether the book inspired him about climate change, Mitic called it a 'tough read' that was bogged down by personal narrative that had 'no relationship to climate change'" (www.cbc.ca/books/canadareads). Chantal Kreviazuk had explained at the beginning of the show that she was speaking from Los Angeles by satellite due to the recent hospitalization of her son for severe asthma. Appearing exhausted and worried, she responded by attacking Mitic personally before recovering with a stronger defense: "This book is the book that we need to read to understand that we are, as human beings, connected to everything around us—each other, God or consciousness, the environment, weather ... and I think that Sheila only proves it through the truth. There's nothing fictional about this, there is no lie needed to be said to tell the truth" (www.cbc.ca/books/canadareads). This did not convince the panel, and though *The Right to Be Cold* was saved on that first day, it was voted off on day three. *Fifteen Dogs*, which had also won the lucrative Giller Prize in 2015, affirmed the central place of the novel in Canada's literary establishment.

As I watched *Canada Reads*, it occurred to me that *Fifteen Dogs*, a novel about dogs granted human consciousness by the gods, provides an excellent illustration of Ghosh's thesis. The creation of anthropomorphized animal characters seems to highlight how "moral adventure" preoccupies the novel as a genre and, as Ghosh argues, blinds human beings to the reality in which we live, a reality demanding attention to the non-human, and to understanding human interaction and impacts on the non-human world. I was also reminded of the calls by Justice Murray Sinclair and others for more education on Indigenous matters that followed the release of the *Truth and Reconciliation Commission* in 2015.[1] The *Canada Reads* discussion lacked the context needed to read Sheila Watt-Cloutier, and in the verbal scrum and takedown that is the format of the show, there was no time to learn or to discover that the history of Inuit writing in English, especially in the form of autobiography, reveals a collective voice that teaches and preserves Indigenous knowledge and that Indigenous knowledge informs all aspects of Inuit life.

TELLING INUIT STORIES

The Right to Be Cold can be read in the context of a distinct form of Inuit storytelling tradition: Inuit autobiography written in English. In keeping with this tradition, Watt-Cloutier presents herself as a spokesperson for her people as a whole, a role in which she seems comfortable, having spent much of her life as an elected leader and advocate. At one time, literary scholars believed that the traditional values of Inuit did not cultivate the sense of an individual self found in autobiography (see Henderson). Despite this perceived cultural conflict, a vibrant tradition of autobiography by Inuit writers dates back to the nineteenth century as Robin McGrath first identified in *Canadian Inuit Literature: The Development of a Tradition*.[2] By the 1960s and 1970s, several landmark works testified to the "abrupt, traumatic change from traditional heritage ways to life in modern Canada" (Lutz in Ulrikah 73). In the autobiographies by Alice Masak French, Mini Aodla Freeman, Lydia Campbell, and Elizabeth Goudie, the autobiographical self is shaped in relation to the community, and in some cases, the author consciously adopts the role of spokesperson representing her people in literary and political terms. Although "not all Inuit writers see their texts as political, even though others may designate them as such" Dale S. Blake shows, "Inuit autobiographers' works do contribute to a political agency that may help to deal with modern-day difficulties" (Blake 5). In her reading of Minnie Aodla Freeman's *Life Among the Qalunaat* and Alice French's *My Name is Masak* and *The Restless Nomad*, Blake perceives Inuit women writing from the "contact zone," both embracing and resisting the cultural expectations of the two worlds juxtaposed in modern life in the Arctic. The "two worlds" stereotype remains strong because of the perception of rapid change in the social and natural environments, and Inuit artists and writers often use it to articulate the unique nature of their identity. Blake explains how individual writers engage with this reality: "French and Freeman refer to customary ways of life, but each must leave traditional environments and adapt to different, non-Inuit conditions" (Blake 128–129). This reality means that the customs McGrath and later Heather Henderson (Blake 62) once observed shaping early autobiography by Inuit women, such as the taboo against speaking about or drawing attention to oneself, have little importance to them. In an interview with Blake, Alice French disavowed the image of herself as a "spokeswoman protesting injustice" (Henderson 65) even though her writing is frank in its criticism of the treatment of Inuit people. Writing about experiences in residential schools and other institutions entails confronting historical and political realities that demand critique, but the writers object to their voices being dismissed as political because this reductive term denies the complex resonances—emotional, epistemological, and historical—in their work. Even though she was working with the heavily edited first edition of *Life Among the Qalunaat*, Blake's analysis remains relevant to the new edition of Mini Freeman's book published in 2015 which includes more of Freeman's original text.[3] As Blake shows, the writers often downplay "political" motives while simultaneously voicing and representing their people and communities, and Freeman's comments in the new edition, particularly with regard to the suppression of her book by the Department of Indian Affairs, display the same sort of reticence. These features can be illustrated by reading two examples of Inuit life writing: *Sanaaq: An Inuit Novel* by Mitiarjuk Nappaaluk and "People of the Good Land" by Alootook Ipellie.

Sanaaq: An Inuit Novel by Mitiarjuk Nappaaluk deals with the period in Inuit history after Canadian colonization from the point of view of Inuit who continued to live on the land. Written in the 1960s, *Sanaaq* became available in English for the first time in 2014 although it was published in Inuktitut in 1983 and translated into French in 2002 by Bernard Saladin d'Anglure, the anthropologist who had encouraged the author through the writing of the text from 1965 until its eventual publication.

Identified as "a novel" in its title and treated as "a series of stories" and "work of fiction" by scholars (Sangster 278), the narrative is structured around daily life, and knowledge of the land is conveyed through details drawn from the author's life.[4] The involvement of Saladin d'Anglure also signals its resemblance to the "as told to autobiography" form produced by collaborations between Indigenous authors and ethnographers throughout the nineteenth and twentieth centuries. In writing *Sanaaq*, Nappaaluk's purpose was to record as many terms in the Inuktitut language as possible for use in language instruction and retention. As a result, a style that has been described as "distinctively straightforward, disarming in its directness" (Sangster 277) is crafted to instruct and to preserve language and story alike. Moreover, Inuit teaching through storytelling is embedded in the narrative.

Several of the book's themes are announced in the first pages, including the journey out on the land and return to the community. The story begins by introducing the main character Sanaaq as she prepares to go out to gather branches: she packs a tumpline, her *ulu*, and a glove for gathering the branches as well as tea, meat, blubber, tobacco, a pipe and chewing tobacco (Nappaaluk 3). Along the way, her efforts are thwarted by her dogs, another theme in the novel, first when they chase off ptarmigan that she wishes to hunt, and next when one of them begins choking on a bone making it necessary to feed it the blubber left from her meal. The blubber dislodges the bone but leaves her to forage for food on her return home. As she walks, she saves the berries she gathers to give to her daughter Qumaq. Once home, she tells her daughter the story of the dog almost choking as she eats the food that has been saved for her. As the mother and daughter talk over the meal, the reader learns that the young girl's father died some time ago while out on the land. Sanaaq uses the opportunity to instruct Qumaq, telling her that before he died, her father wished for her to behave herself. This opening passage presents an introduction to the modern Inuit world depicted in the rest of the book. As Bernard Saladin d'Anglure points out in the foreword, this period was characterized by a return to a subsistence lifestyle brought about by the closure of two trading posts in the area after the collapse in the price of Arctic fox furs during the Depression and the Second World War. With the Inuit economy that had developed around trade with the Qalunaat in decline, families with the means to do so relocated to areas that still had trading posts, leaving the others to resume a hunting and gathering lifestyle. Saladin d'Anglure borrows the term "false archaism" from Lévi-Strauss to describe this life. In this "early transitional period" from the 1930s to 1950s, *qalunaat* authorities became ever more present and powerful.

The intrusion of southern authority becomes the source of marital strife leading to domestic violence when Sanaaq's husband Qalingu tries to send their son south for medical treatment. Her resistance to sending her son away for medical treatment documents the increasing presence of southern institutions in Inuit daily life. In a panic, she takes the child and runs away but soon realizes that they will not make it alone on the land. She returns the suffering child to the community where he is looked after while she goes to find Qalingu, who is also looking for her. When he finds her, believing her flight has led to the child's death, he beats her so severely that she too must go away for treatment. When the police warn Qalingu that if he beats her again he will go to prison for five years, it shows the authority the police have. This is the time when the Royal Canadian Mounted Police (RCMP) virtually governed the North, making often devastating decisions such as the cull of huskies that took place in the 1950s and 1960s.[5] Significantly, Sanaaq's journey south and convalescence in a southern hospital are described with very little detail and the southern places are never described as part of the setting.

The focus on the subsistence life casts Modernity as an outside force acting on the community. In other scenes, the outside world is threatening, as in the passage that describes how the young girls are startled by an airplane:

> While [Aanikallak] and Qumaq amused themselves, each of them building an igloo, they suddenly heard a large plane and were very frightened by the noise it made … They scampered off to their homes. Qumaq tripped and fell several times, so overcome with fright was she. Her mother asked, 'What's with you? *Ii! Aalummi!*' and she ran out to see. 'Look at the big plane! *Qatannguuk!* The plane disappeared. The children were still very frightened." (Nappaaluk 91)

As the northern regions are gradually colonized, aviation was "a new face of imperialism" extending imperial expansion across both space and time (Vance 157). Even though the children already know what the plane is—for they know the word to use to tell Sanaaq what they have seen—they are depicted as surprised and terrified. As the characters in the book carry on with hunting and gathering, the modern world is all around them. The lasting changes to the environment have not yet occurred, and the traditional life on the land continues despite the rapidly increasing southern presence in daily life. The characters may rely on trading with the *qalunaat*, as they had for a generation or so, but they resist the authority of the Canadian state over their affairs, especially the *qalunaat* institutions that are gaining ground in the community, by focusing on daily life in the community and by refusing to narrate the characters' experiences when they travel outside.

As in the first scene, Qumaq's youth and inexperience are signified by her clumsiness when she sees the plane, though this time it is fear not excitement that causes her to fall down. Qumaq's character, particularly her need for her mother's constant vigilance and guidance, serves as a literary device that focuses on Sanaaq's role as a teacher and elder. In this context, Qumaq is a surrogate for a reading audience that also needs instruction in Inuit ways of doing things, and whose education is an underlying purpose of the book as a whole. Qalingu also highlights teaching through oral tradition in his approach to problem solving. After seeking out Taqriasuk with questions, Qalingu states: "Thank you! I won't forget any of what you told me and which I did not know before. I need to be taught ... Without elders the Inuit are nothing for there is much knowledge that the elders alone possess" (Nappaaluk 112). To which the elder replies: "My knowledge comes not from me but from my ancestors. It seems to be mine but, in fact, it comes from people who preceded me. I pass it on to you, to all of your descendants and all of your kinfolk" (Nappaaluk 112). In contrast to the scene in which Qalingu acts alone and makes the indefensible decision to beat Sanaaq, this scene illustrates the appropriate way to acquire knowledge and to make decisions.

Sanaaq is an important source of Indigenous Knowledge, written on the cusp of the permanent changes that would come to the Arctic. Indeed, like many works of Inuit literature, it defies generic classification, presenting an autobiographical depiction of everyday life that is layered with meaning, preserving both the Inuktitut language and collective memory in written form, and depicting life in Nunavik as the Inuit were making the transition from nomadic to settled life. The same history is recalled in Alootook Ipellie's essay "People of the Good Land," an essay that opens with a passage from John Rae:

> If the moralist is inclined to speculate on the nature and distribution of happiness in this world [let him consider the Eskimo]: a horde so small, so secluded, occupying so apparently helpless a country, so barren, so wild and so repulsive; and yet enjoying the most perfect vigour, the most well fed health. (Rae, qtd in Ipellie 19)

In the paragraphs that follow, Ipellie rewrites and parodies Rae's words: a "modern-day explorer can use the same words and still be close to explaining the truth about the nature of the Inuit and his environment," Ipellie notes, but "[i]f today's moralist is to speculate about the nature and happiness of the present-day Inuit, he is bound to be a little surprised by what history has done to them ever since that day in 1830" (Ipellie 19). Ipellie moves quickly from the Arctic explorer's outsider position and perspective to the objective view of "the present-day Inuit," and continues in the third person as he outlines important aspects of Inuit culture and history. What has happened to "them," and what makes "Inuit unique among cultures in this world" is "their" ability to thrive in the Arctic: "Even though they live in one of the world's most inhospitable climates, their enduring characteristic is a people warm of heart who are always ready to help anyone struggling with life" (Ipellie 19).

While reading the content of the essay offers valuable insight, it is also important to consider the literary techniques used for, as Keavy Martin puts it, "[r]eading Indigenous autobiographies as literary texts ... aims to liberate Indigenous authors from ethnocentric assumptions regarding their choice of genre and from ethnographic readings that diminish readerly appreciation of their skill" (105). For the first part of the essay, Ipellie takes the voice of the outsider, the third-person perspective used by explorers, anthropologists, and others who study Inuit subjects in order to trace the colonization of his homeland from the point of view of one who has lived through it. By this masterful rhetorical technique, he introduces and affirms the importance of self-representation in Inuit writing. He describes the traditional life of the Inuit, the relationship to their environment, and the role of oral tradition in their survival, then moves to the arrival of Europeans, recounting John Cabot's kidnapping of three Inuit who were put on display in Europe. In this story, he draws the reader's attention to the feelings of the captives who died of the diseases they encountered. For Cabot, they were "like brute beasts" who "spoke such speech that no man [in Europe] could understand them" (qtd in Ipellie 20–21), but Ipellie imagines how they must have longed for home though no one recorded their feelings. The captives' helplessness foreshadows the "cultural whiteout" that Inuit experienced when new powers asserted themselves in the Arctic as "[t]hese colonizers [the US and Canada] decided that the nomadic Inuit were ripe to be assimilated to the larger, dominant societies" (Ipellie 22). It is at this point in the history that Ipellie inserts himself, using the first-person pronoun. Likening the incursion of outside governments, which he personifies as an Orwellian "Big Brother" and its counterpart "Thought Control" plowing like bulldozers through the Arctic, he notes "I, too, felt the thrust" (Ipellie 23). Caught up in a sudden change from living on the land to living in newly created settlements, Ipellie shows that the Inuit were not fully aware: "I have to admit to being blind to what was happening to us" (Ipellie 23). He also admits the allure of *Qulunaat* goods: "I too was attracted to their colourful world, their sweet food, and the ingenuity with which they had invented toys for the modern world" (Ipellie 23), highlighting the material conditions created by colonization and the lasting role of trade in that encounter.

The famous photo of Niviatsianaq working at her sewing machine is reproduced at this point in the essay. Niviatsianaq, or "Shoofly," the woman who so fascinates Kathleen Winter in *Boundless*, provided clothing and supplies to American whalers while overseeing the work of other women in the community. The photo of Niviatsianaq accompanies several photos of Inuit including one of two girls wearing bonnets and dresses made in the late Victorian style. It documents the long history of women's labor that sustained Arctic exploration and colonization and, like the photo from *Nanook of the North* in which Nanook encounters a gramophone for what seems to be the first time, it simultaneously stages the arrival of Modernity and its technologies. In Ipellie's essay, the photo of Niviatsianaq and her sewing machine accompanies the part of the essay describing "one of the most vivid memories" of his childhood, the story of "a woman who had been found totally naked, half buried in the snow" (Ipellie 24–25). Ipellie recalls his child self not knowing "how to react to this rumour" because she was not a member of his immediate family and innocently wondering how the woman met her death, whether gasping for her life as a storm raged or closing her eyes in cold-induced sleep (Ipellie 25). In retrospect, he knows the "real cause of her death that was most telling about the dilemma the Inuit in our community were facing: alcohol abuse and its related problems" (Ipellie 25), and his oblique admission that he "felt as helpless as the next person," and was also "one of its victims" (Ipellie 25) is a poignant moment foretelling his own untimely death in 2007. As Kenn Harper's touching tribute tells, Ipellie's death from a heart attack at the young age of 56 ended years of personal struggle with the alcoholism that led to estrangement from friends and family, isolation, and periods of homelessness. While acknowledging the hard truths of his life, the obituaries by Harper and others do not dwell on them but instead remember Ipellie's incredible talent and originality as an artist. In "People of the Good Land," Ipellie highlights the genius Inuit have displayed for artistic and technological innovation, a genius his work shares.

Writing before the publication of "People of the Good Land," Dale S. Blake presents Ipellie's art and writing as a performance of autobiography that defies and remakes the genre. The "self" in his work is an externalized character who displays characteristics of both the author and the fantastic heroes of Inuit stories. In cartoons, essays, and poems, and especially through the creation of the self as a shaman figure in *Arctic Dreams and Nightmares*, "Ipellie, working alone, reinvents himself as a disembodied subject, freed from the limitations of mortality and the physical self. At the same time, he speaks for his people against colonization and disempowerment" (Blake 166). What Blake calls Ipellie's "super-self," a "malleable character who still retains his 'Inuitness'," works to counteract the representation (and marketing) of Inuit art as anonymous creations deemed more authentic for subordinating individual artistry to cultural identity (Blake 170). She argues that, through political satire, especially targeting current events like the European ban on seal products, Ipellie "reaffirms his connection to his people" taking up important issues in Inuit life across the Arctic (Blake 188). By telling the story of his early life in "People of the Good Land," Ipellie traced the incursion of the Canadian government into Inuit lives beginning with movement of Inuit off the land into settlements. A sharp observer of the effects of colonization, Ipellie produced a large body of satirical drawings and stories as well as journalism and poetry. His pen-and-ink drawings in publications like *Kivioq*, which he also edited, his exhibitions, and his major work, *Arctic Dreams and Nightmares*, all commented on the political and social reality of his homeland while drawing on traditional stories and knowledge. In the conclusion of his essay, he returned to the subject of the epigraph with a section heading, "Today's Arctic is Alive and Well":

> While it is resource-rich, its ecology is one of the most fragile in the world. The land, the animals, and the Inuit are inseparable and their relationship seems to have been made in heaven. This is a reality that has never been properly understood by the dominant society. For this reason, Inuit have much to teach others in other lands about the land's culture and heritage. (Ipellie 31)

Indeed, his essay claims the authoritative voice, first by deconstructing the explorer's observations, then by taking the objective perspective of the Arctic expert before finally replacing these perspectives with a first-person account of lived experience grounded in Indigenous Knowledge: "We Inuit refer to our homeland as 'Paradise on Earth,' and we know what we are speaking of" (Ipellie 31).

When Ipellie describes "the healing process, of reclaiming our traditions and heritage," he sees it in relation to people across the Arctic: "It's the same thing, the same feeling we have in all circumpolar communities, like Alaska and Greenland, northern Canada, but slowly in Siberia also" (qtd in Blake 168). "People of the Good Land" stands as witness to what Ipellie calls the "cultural whiteout" of the 1950s and 1960s that he likens to being in a whiteout "trapped and unable to go forward since you could not see clearly where you were heading" (Ipellie 26). Emerging from the cultural whiteout, Inuit writers tell their stories, stories that are the teachings that ground Inuit life. Reading these classics of Inuit literature provides a context in which to place Watt-Cloutier's memoir. The importance of the autobiographical voice as an expression of the collective life of the Inuit is central to the story the writer tells about learning and preserving the knowledge necessary for life. This is what *CBC Reads* missed: every detail of Watt-Cloutier's life story is also a call for action on global warming.

READING *THE RIGHT TO BE COLD*

The first glimpse of the Arctic in *The Right to Be Cold* is from the land. "The world I was born into has changed forever" writes Sheila Watt-Cloutier:

> For the first ten years of my life, I travelled only by dog team. As the youngest child of four on our family and ice-fishing trips, I would be snuggled into warm blankets and fur in a box tied safely on top of a *qamutiik*, the dogsled. I would view the vast expanses of Arctic sky and feel the crunching of the snow and the ice below me as our dogs, led by my brothers Charlie and Elijah, carried us safely across the frozen land. I remember just as vividly the Arctic summer scenes that slipped by as I sat in the canoe on the way to our hunting and fishing grounds. The world was blue and white and rocky, and defined by the things that had an immediate bearing on us—the people who helped and cared for us, the dogs that gave us their strength, the water and land that nurtured us. (Watt-Cloutier vii)

In the author's memory, the land the sled travels over is solid and stable. In the present of her remembering, it is anything but; for in the intervening years, as Watt-Cloutier narrates, the Arctic has changed rapidly. Watt-Cloutier was born in 1953, after the Americans left the airforce base in New Fort Chimo and the community gradually moved from Old Fort Chimo where the Hudson's Bay Company (HBC) post was located, following the construction of a new school there, the relocation of the Anglican and Catholic missions, and the addition of a nursing station. As "one of the rare Inuit of that era who was proficient enough in English to be considered bilingual," her mother, Daisy, worked as an interpreter for the staff of the nursing station and the health workers who flew in to the community. When Sheila was seven, her mother built a new house for the family with the lumber from a dismantled power station (Watt-Cloutier 7). She recalls that moving into the house "filled the whole family with a sense of pride and accomplishment" (Watt-Cloutier 8). While her mother was the one who provided for the family, she also credits the "calm and loving presence" of her grandmother, who cared for her and her siblings during the day, for creating "a family that was strong and dignified and that had moved beyond their initial struggles in the rapidly transitioning culture" (Watt-Cloutier 11).

When Watt-Cloutier was only ten years old, she was sent to Blanche, Nova Scotia to go to school. She had been selected for the federal government-sponsored program as one of the "promising" children who would be educated in the south. Not only was the flight south traumatic, but the family who billeted her had no understanding of Inuit culture. She especially missed the country food that for Inuit nourishes both body and spirit. Not only the food but the way of life was alien to her, and the homesickness she felt was intense. When her host parents decided to censor her letters home, she felt profoundly betrayed and alone in the world. What she later calls her "first wounding" had an effect on her confidence, silencing her for many years. Later, at residential school in Churchill, Manitoba, she felt less alone because, unlike most residential schools, the school allowed siblings to be together, and the discipline she applied to her academic studies reminded her of the rigors of Inuit traditional life. With a characteristically matter-of-fact tone and precise language, she narrates the degrading treatment and indignity suffered by the children that is so familiar to readers of residential school memoirs:

> We landed in Churchill in the middle of the night, tired from the long trip. We were greeted by supervisors, the residential school "parents" who would look after us in the dormitories in shifts. They told us to line up immediately.

They gave us each a number and a set of clothing—underwear, T-shirts, shirts, pants, socks and so on. I was number eight. Then they instructed us to get ready to hop into the showers. (Watt-Cloutier 38)

The children were then treated for lice and marched to the dormitories:

Remember, we were young children and teens, arriving in the middle of the night in a strange setting ... There was no dignity in this introduction to our new home, not to mention a blatant absence of welcoming, loving energy. And while the presence of my sister and the other Inuit children made me feel a whole lot safer than I might have otherwise, my introduction to this new institution was still an unsettling one. (Watt-Cloutier 38–39)

Watt-Cloutier's gentle way of describing what must have been harrowing for the children reflects her prose style throughout the book. In exact and measured language, she describes the deep hurt caused by the system and the widespread suffering caused by residential schools, and she notes how a "culture of silence surrounded this violation of children" (Watt-Cloutier 46). Though she herself did not witness or experience physical or sexual abuse, others did, and as she says, "even one abused child is too many" (Watt-Cloutier 48). Referencing the *Truth and Reconciliation Commission*'s work, she emphasizes the depth and extent of the damage done:

I've lived most of my adult life coming to terms with my own intergenerational legacies, including being uprooted as a young child and losing my language and my connection to my culture, to my country food and to my family, but even this doesn't compare to the horror that Aboriginal children went through in the mission-run schools during that time. (Watt-Cloutier 47)

Although she details the personal aspects of her education at Churchill and points to the other leaders, including Jose Kusugak, Eric Tagoona, Mark R. Gordon, and John Amagoalik as well as other artists and teachers who attended, she explains that the cultural impact of residential school was damaging for everyone. Many of the skills taught at the school were similar to those that the children would have learned at home, but the method of teaching was so different that it was clear to Watt-Cloutier that "We were being deprogrammed from our Inuit culture and reprogrammed for the southern world" (39). The differences in teaching methods would lead her to reflect on the role of education as she embarked on her own work in her community:

Our future, in many ways, lay in what we knew from the past but had mislaid. In Inuit culture, and, in fact, in all Aboriginal cultures, our elders were the source of wisdom, holding a long-term view of the cycles and changes of life. Wisdom was forged through the independent judgment, initiative and skill required to live on the land and ice. The land was our teacher, and our hunters knew the value of patience and trust. Without them, we would perish. (Watt-Cloutier 112)

After working in health care, Watt-Cloutier took a job with the Kativik School Board in Kuujjuaq. She describes this time as her introduction to the institutional and political arenas in which she would spend her career. Working as a guidance counselor, she gained a better understanding of the challenges facing Inuit youth and "how the wounding of the previous generation was having a dire impact on the next generation" (Watt-Cloutier 81). She became an outspoken critic of the school system, eventually resigning altogether, and although the school board opposed her inclusion in the Nunavut Education Task Force, Johnny Adams invited her onboard the committee which included other Inuit leaders Minnie Gray, Mary Simon, Annie Tulugak, Josepi Padlayat, and Jobie Epoo. The title of their final report, *Silatunirmut: The Pathway to Wisdom*, made explicit the connection of education reform to broader social and cultural issues. Children in Inuit communities were not being challenged to learn as traditional methods of teaching survival on the land and ice had once demanded. Instead, their developing desire for autonomy was being frustrated within the existing school system which served as a "reflection of the growth of many other dependency-producing institutions in our Aboriginal communities" (Watt-Cloutier 111). The more governments tried to "fix" social problems, the more damage was being done to the communities as the traditional role of the elders was usurped by new authority figures. "Even institutions as seemingly benign as social and health services and the courts had been deeply corrosive to a culture encouraged to rely on themselves for directions and acquiring wisdom" (Watt-Cloutier 111–112). To regain sovereignty, reform of the school system would be necessary to restore the social structure of Inuit society broken down by colonization.

The experience of working on the Kativik School Board exposed Watt-Cloutier to the "frustrating and painful experiences" of political life. It also helped her overcome what she calls her "first wounding—the silencing of my voice in Blanche" (Watt-Cloutier 97). In the chapter entitled

"Finding Our Voice," she tells how this moment led to her eventual role as a spokesperson for her community and the Arctic as a whole. Watt-Cloutier's life spans the decades of social change followed by rapid climate change, and while certain practices and aspects of traditional culture have altered, much has also remained the same. The way of life she knew as a child may have changed, but adapting to global circumstances, the culture has retained the values and lessons of Inuit ways of knowing and being, making her a knowledgeable speaker on Inuit tradition. In traditional hunting society, Inuit learned to read the land, and this practice became a social way of being:

> Inuit have always placed a high value on the ability to be calm, controlled, focused and reflective ... Even physically, we Inuit have learned the importance of quiet: the ability to remain still is an essential survival skill on a hunt. These habits have become part of our social behaviour, too. (Watt-Cloutier 12)

"Calm, controlled, focused and reflective" are also good adjectives to describe the text of this memoir which never wavers in its objective to persuade readers that all human beings have a *right* to a stable climate and healthy environment.

After many years of traveling and living on the road, Watt-Cloutier felt she "needed to spend some time in the Arctic, living among my fellow Inuit" (Watt-Cloutier 181), so after completing the work on the Stockholm Convention on Persistent Organic Pollutants, she decided to move to Iqaluit because, as she puts it, "when I came home, I wanted to be *home*" (Watt-Cloutier 181). With her daughter and grandson living in Iqaluit at the time, Watt-Cloutier immediately reconnected with her family and began to reflect on her identity: "I started to sense at a deeper level that representing my fellow Inuit and the North, fighting to preserve our way of life, fortified my sense of self. I was Inuk—heart, mind, and soul. And I knew deep in my bones that the only place that was truly home was the Arctic" (Watt-Cloutier 183). While living in Iqaluit, she was able to center herself and to prepare herself for "the next chapter"—her "fight for the right to be cold" (184). It is in this section of the memoir that she relates the touching story of watching her adult son Eric, a commercial pilot, fly over her home for the first time.

It is a significant moment because, in the North, the airplane was an agent of the rapid social change: it brought white settlers and government agencies into the North who used it to relocate northern populations and to remove Indigenous children to residential schools. With the arrival of the Americans during the Second World War, airplanes became commonplace in Nunavik, illustrating the coexistence of modern and traditional life in the Arctic. When her son calls to say he will fly over Iqaluit at a specific time, she waits on the deck of her house until she spots the lights of his plane overhead. It is a symbolic homecoming for Watt-Cloutier who first traveled by airplane from Kujjuaq (then called Fort Chimo) to Roberval, Quebec to have her tonsils removed when she was eight or nine (Watt-Cloutier 23). She recalls returning home after the month away: "When we landed in Fort Chimo, all the mothers gathered at the airstrip to meet us. We were so excited to see familiar faces and hear everyone speaking our language again" (Watt-Cloutier 25). A few years later, she would leave home on a plane bound for Nova Scotia, and later, an airplane would take her to residential school in Churchill, Manitoba. When President Kennedy was shot, the children, having been told that he was a leader who prevented wars, expected the worst: "We fearfully looked up to the Arctic sky, waiting for the airplanes, the fighter jets, to arrive and for war to start" (Watt-Cloutier 21). Like the children in *Sanaaq*, who ran from the sound of an airplane overhead, Watt-Cloutier and her friends, living near an airforce base, expected a threat from the sky. This response was typical everywhere after the attack captured in Picasso's depiction of the bombing of Guernica as throughout the world "[t]he airplane, so recently seen as the pinnacle of human achievement, was evolving into a sinister killing machine capable of spreading destruction. The sound of an airplane engine had once sent people dashing outside to catch a glimpse of the aerial wonder; now it was as likely to send them scurrying for shelter" (Vance 221). These scenes therefore connect the children in Nunavik to a modern world in which conflicts are global, but the more pressing threat they pose is as vehicles of colonialism transporting Inuit out of their homeland into the institutions of the south. Having lived through such displacements as a child, Watt-Cloutier becomes an Inuit leader who travels the world by plane, living the coexistence of modern and traditional life that her son's flight overhead represents:

> I looked up to the sky. It was a cool fall night, the inky sky full of stars. At precisely 12:09 the stillness of the northern sky was broken. I heard the distant roar of a jet and saw the flashing lights above me. It was my boy. After everything Eric and I had been through, after all his painful struggles in school, there he was, doing what he had always dreamed of—he was literally soaring. I put my hand over heart and felt my throat get tight, my eyes fill with tears. (Watt-Cloutier 183–184)

In this scene, the reader is reminded that, though Watt-Cloutier has returned north to get respite from the hectic globe-trotting life of an activist, she has not retreated from the modern world. The Arctic is as much a part of the modern world as the places she has visited.

By the age of ten, Watt-Cloutier had made her second trip south by airplane, and her career as an advocate for the people of the Arctic would eventually take her, by plane, all over the world. Eventually, seeing her homeland from the air becomes as familiar and comforting as the view from under the blankets of the *qamutiik* in her memory: "Every time I fly home from far-flung places I can't help but smile as a northern town comes into focus against the blue-tinged vastness of ice, snow and sky" (Watt-Cloutier xv). The Arctic sublime evoked in literary representations of Arctic flight in this passage resonates in the reader's imagination with the vast array of Arctic images circulating in contemporary culture. But Watt-Cloutier wants the reader to see that the Arctic is "more than polar bears and seals" (Watt-Cloutier xv): what she sees as she looks out her window are welcoming communities full of Ipellie's "people who are warm of heart" and whose survival in places of warmth and life depends on remaining cold. This may seem paradoxical to the outsider, but as Watt-Cloutier shows, it is beautifully consistent from an Inuit point of view which she offers by telling the story of her early life before embarking on the account of her years spent negotiating at the international level.

The opening scene of comfort and contentment from beneath the blankets of the *qamutiik* depicts a lifestyle that was violently interrupted by the killing of the sled dogs also described in *Sanaaq*. The destruction of the sled dogs, like the forced relocation of Inuit communities to remote parts of the Arctic such as Grise Fjord, exemplifies the control of the Canadian state over the lives of the Inuit. Watt-Cloutier recounts the Inuit testimony describing how the RCMP officers and government officials were authorized to cull the sled dogs belonging to Inuit hunters across the Baffin Region despite the fact that the "importance of the sled dogs to Inuit, in particular our hunters, can't be overstated" and losing their dogs was a blow from which many hunters and their families never recovered. By this single policy, the Canadian government disrupted Inuit social life, and "many now suspect that the destruction of the dog teams was another way to force Inuit families to move from outpost camps into settlements by removing their only mode of transportation" (Watt-Cloutier 71). In other words, the cull, ostensibly conducted to rid the communities of canine distemper, was a means to an end, an explanation that is supported by Inuit testimony that dog teams were "not inspected for illness, no questions were posed about their health or behaviour" before they were rounded up and killed or shot as they stood in harness (Watt-Cloutier 71). The impact of this loss on the transmission of Inuit knowledge of the land was devastating. Nevertheless, the gradual separation of Inuit from their knowledge base through such policies did not destroy the Indigenous knowledge passed on through tradition which remains ever present in modern Inuit life.

As she cherishes memories from the safety of the *qamutik* and feels the loss incurred by the cull, Watt-Cloutier recounts being born into a community that was both traditional and modern joined to the world by modern communication and transportation technologies. With the former American Air Force base nearby, Arctic aviation was always part of her world, a world in which modern and Inuit technology existed side by side. As a child, she remembers waking to the sound of her grandmother's radio broadcasting voices coming from Greenland. As in many northern communities, the once nomadic Inuit were forced to settle around trading posts, missions, and schools. Watt-Cloutier's personal account of contact and colonization begins with the story of generations of resourceful Inuit women living, and even thriving, outside the traditional hunting culture of their people. Her grandmother had three children with a Scottish (HBC) clerk named William Watt. When he was transferred in 1926, she remained with her children in Kuujjuaq, then known as (Old) Fort Chimo. Without a hunter in the household, the family had to stay close to the HBC post where her grandmother could work as a domestic. Even so, Watt-Cloutier's grandmother was forced to give her younger daughter Penina to another family to be raised. Her eldest daughter Daisy had three children: Charlie, who would become a Senator, Bridget, and Sheila, and together, they adopted Elijah. The transition from a trading post to a strategic base on the Crimson Route connecting the US Air Force to their allies in Great Britain during the Second World War, and with it the colonization of Nunavik, happened overnight (Watt-Cloutier 5–6).

At home in the Arctic, Watt-Cloutier reflected on her past and renewed her energies. Speaking to the Arnait Nipingit Women's Leadership Summit in 2010, she explained that the style of leadership derived from Inuit social habits is one of many lessons of her upbringing in Nunavik. This kind of leadership means "'never losing sight of the fact that the issues at hand are so much bigger than you'" and "'working from a principled and ethical place within yourself'" to "'model, authentically, for others, a sense of calm, clarity and focus'" (Watt-Cloutier 268). Her success as a leader offers a model of how modern Inuit life can teach others to succeed. In both public and professional spheres, she has worked tirelessly to shift the discussion of climate change "out of the realm of dry economic and technical debate" (Watt-Cloutier xii). Her commitment to this goal is deeply rooted in Inuit life and culture, making the story of the author's life inseparable from the work she does. As Jackie Price explains, "[a]ll leadership roles support the community in their own way," with support from other knowledgeable and experienced members of the Inuit community, "a leader's authority is upheld for as long as the community continues to respect and recognize the leader, and a leader's direction is respected as long as it falls within the leader's area of expertise" (Price 133). In these terms, Watt-Cloutier continues to play a crucial role as an experienced negotiator and communicator in the movement for Inuit sovereignty. As Ipellie would say, she knows what she is speaking of—and what she knows is transforming the language of ecology and human rights.

In addition to the patience and silence that are cultural values in Inuit society, what I would call a principled pragmatism learned from Arctic life permeates Watt-Cloutier's narrative. This approach is evident in her search for concrete results, her preference for persuasion, and her dedication to negotiated solutions. The position brings her into conflict with members of communities wishing to pursue resource development and to welcome extractive industries on their land. For example, she describes her dismay during the annual meeting of United Nations Framework Convention on Climate Change (UNFCCC) held in Copenhagen in 2008 at the decision of representatives from Alaska to argue for an exemption from emission mitigation and the news that officials from Greenland were considering taking the same stance. Speaking after learning of this ironic turn of events, Watt-Cloutier pleaded with those attending the Indigenous People's Day at the gathering to hold firm to their own "moral compass" and to "resist the urge to compromise those very values by adopting quick fixes to our economic and social problems" (296). The influence Inuit have, she continued, "springs from our ethical authority, and if we lose that moral high ground, we lose our influence" (Watt-Cloutier 297). This exchange demonstrates how Inuit societies craft a position that is in line with their values while managing competing interests. As Ann Fienup-Riordan observed in her study of Yup'ik society, Inuit are subject to a harmful double standard: stereotyped as naturally "peaceful" and "ecological" people, they are blamed for failing to live up to the ideal when using resources (see Fienup-Riordan 2003 *passim*). The stereotype of the innocent, passive Inuit has a long history in western representation dating back to Samuel Hearne's depiction of the "massacre" at Bloody Falls (see also McGrath 1993 and Cameron 2015). Watt-Cloutier is mindful of the trap and careful to distinguish a pragmatic position derived from Inuit knowledge and values from the stereotype of the smiling, peaceful Inuit that has been used to justify control over their lives.

UNDERSTANDING *SILA*

Before cruise ships plied the increasingly ice-free waters of the Northwest Passage, and aviation opened northern regions to outsiders, before ships came looking for passage and lost explorers, Indigenous peoples traversed the Arctic by foot, sled, kayak, and canoe. Based on experience and tradition, their knowledge of land, sea, and ice is ancient and adaptable. Inuit living on the land were the first to observe the effects of global warming, and working in collaboration with scientists, they are continuing to map climate change. As Watt-Cloutier explains: "*Silatuniq* is the Inuktitut work for wisdom—and much of it is taught through the experiential observation of the hunt. The Arctic is not an easy place to stay alive if one has not mastered the life skills passed down from generation to generation" (Watt-Cloutier ix). While some early explorers like Franklin were ignorant of Indigenous knowledge, and Inuit eyewitnesses were slandered in the British press, other European travelers, John Rae, Roald Amundsen, and Knud Rasmussen among others, believed in the people they met and learned to embrace Indigenous knowledge and technology. Contemporary environmentalists are also learning from the Inuit term *Sila* (translated variously as "weather," "air," and "breath" but also containing the idea of wisdom), a force infusing the Arctic world with life that captures the interdependence of human and natural world in the Anthropocene.

While the Inuktitut word *Sila* has been adopted by climate change researchers, Timothy B. Leduc argues that translating the word narrowly as "climate" or "weather" does not do justice to the full range of its meanings (21–26; see also Cameron). The literal translation of Inuktitut words into English terms often fails to translate the cultural context in which the Inuktitut words have meaning. In order to fit the term into climate change discourse, it has also been recast. This appropriation serves to reaffirm the dominant ways of thinking about the problems and solutions associated with climate change because it is "the translation of climate change as a wholly environmental phenomenon and not as a matter of social, political, or structural injustice and harm that naturalizes adaptation and resilience as appropriate responses to climatic change" (Cameron 279). In environmental discourse, the concepts generated to talk about saving the Arctic environment, concepts such as "adaptation" and "resilience," could become measures by which outsiders judge northern societies, especially as they grapple with changes to the environment. The ethnographic evidence contained in Rasmussen's accounts of his conversations with Inuit shamans about *Sila* that LeDuc examines suggests that Inuit are speaking about much more when they use the word (21–32); indeed, modern Inuit may also be referring to the "cultural and spiritual dimensions that interpret climate change as the world's ethical response to improper human actions" (Cameron 242).

While the global environmental crisis calls for learning about and from Indigenous traditions, it also requires caution. In *Staying with the Trouble: Making Kin in the Chthulucene,* for example, Donna J. Haraway uses Inuit examples to help think through a material semiotics that is "always situated" yet responsive to global issues. Haraway is drawn to the traditions shaping what she calls Tanya Tagaq's "utterly contemporary" Polaris Music Prize performances from her 2014 album *Animism*: "Hunting, eating, living-with, dying-with, and moving-within the turbulent folds and eddies of a situated earth: these were the affirmations and controversies of Tagaq's singing and website texts and interviews ... She wore seal fur cuffs during her Polaris performance; she affirmed the natural world and hunting by her people" (Haraway 165). As the title of an album, *Animism* evokes the themes Tagaq explores as an artist, but outsiders should be careful not to attribute particular spiritual beliefs to Inuit people generally. Haraway's goal is not to explain the complex worldview underlying Tagaq's representation of humans and animals but to imagine possible worlds in which the interdependence of species is as meaningful to everyone as it is to Inuit. Nevertheless, while valuing the "situated" nature of the performance, the rest of the argument tends to subsume the situated or the particular, collapsing contexts as when "Madagascar, the Inuit Arctic, and the Navajo-Hopi Black Mesa" are folded into the category of geo-political "zones" as objects in need of ecological care (Haraway 202n79). As provocative as Haraway's language and imaginative arrangements of ideas are, theoretical abstraction can also marginalize and distract from the particular knowledge and specific lessons Inuit writers bring to understanding the global crisis.

"The Anthropocene has reversed the temporal order of modernity," writes Amitav Ghosh, "those at the margins are now the first to experience the future that awaits all of us" (62–63). If we consider Modernity to be produced in a center and exported to the periphery, this statement makes sense; but Arctic peoples have always been part of generating Modernity and continue to contribute to its unfolding. For environmentalists like John McCannon, Arctic lives are often seen as "lived on the margin" in the sense Ghosh articulates, where "weather and scarcity lend an extra intensity to the struggle for existence there" (McCannon 2012, 103). From a circumpolar perspective, of course, the Arctic is not peripheral; it is central. In *The Last Imaginary Place*, Robert McGhee describes being in the Arctic and "the overwhelming sense that I was not standing at the distant margin of a world, the end of the earth, as far as one could travel from Europe. Instead, I was standing at the very nexus that for millennia has linked the peoples and cultures of Asia and America. It was a world in which many nations and cultures had flourished, among them the Inuit and their way of life" (129). Despite the longevity of Inuit habitation in the Arctic, the "unpredictability of weather in recent years devalues the knowledge that is the most basic resource of Arctic hunters and travellers" (McGhee 267). Frank Sejersen and Michael Bravo highlight the way discourses of Modernity cast traditional societies as other, and in recent discussions, equate Indigenous peoples with an environment that needs saving: "Global climate change constitutes a new platform from which indigenous people can form, mobilize and articulate different concerns, but they run the risk of becoming ensnared in the discourse of marginality" (Sejersen 23; see also Bravo 2000, 2009). From the outside, Sejersen warns that the "agency of Arctic Indigenous peoples is evolving and being shaped by climate change narratives," and he argues, "the very focus on vulnerability and impact as the organizing device generates certain assumptions about indigenous agency and about the resources that enable them to deal with the consequences" (Sejersen 23). Knowledge production from the Arctic, such as Zacharias Kunuk's film *Qapirangajuq: Inuit Knowledge and Climate Change*, focuses on agency and sovereignty, not the marginality of Inuit people.

In a chapter devoted to "The Voices of the Hunters," Watt-Cloutier describes the growing science based on collaboration between scientific and Indigenous Knowledge teams as exemplified in the Arctic Climate Impact Assessment project funded by the U.S. National Science Foundation, the National Oceanic and Atmospheric Administration, and the University of Alaska. At the center of this research were hunters' observations of thinning ice, disappearing sea ice, and stickier, softer snow. These changes were making it increasingly difficult to live on the land: rotting ice made sealing and whaling dangerous; a "lack of crisp, dry snow made the running of sleds and snow machines much slower" (Watt-Cloutier 188); and the "deep, dense snow required for building snow houses was extremely hard to find" (Watt-Cloutier 189). By recounting what the hunters discover out on the land, Watt-Cloutier demonstrates how the expertise of the hunters bears the same weight as the scientific expertise of outsiders in this knowledge base while revealing her own expertise as a public servant, elected representative, and negotiator in her able handling of evidence. In the text, quotations from her speeches and position papers serve as evidence of her ongoing commitment to discussion and persuasion through carefully crafted rhetoric supported by concrete evidence. This rhetorical structure is consistent with her argument that knowledge should form the basis of decision making. With each example drawn from her experience, Watt-Cloutier confirms her credibility as a spokesperson and thus supports her argument.

Inuit hunters also noticed changes in the animals: seals and caribou were fatter; lemmings, a principal food source for larger animals and birds, were disappearing; and fish with softer flesh meant not only that the food supply became precarious but that it was less nutritious. "But perhaps most shocking of all," writes Watt-Cloutier, "was that the very ground beneath our feet was no longer solid" (191). The melting of the permafrost means travel is difficult and caching meat next to impossible. It also accelerates coastal erosion as the Inupiaq people of Kivalina, Alaska know well. As communities around the world like Kivalina were suffering and the people of Pangnirtung were watching their river change course (see *Qapirangajuq*), some continued to debate the existence of global warming. Canadian Prime Minister Stephen Harper characterized the science of climate change as "tentative" and "contradictory," claiming the jury was still out (Watt-Cloutier 196). "But," Watt-Cloutier writes, "no one in the Arctic could possibly have believed the jury was still out" (Watt-Cloutier 197). This contrast demonstrates the gulf between those living in the Arctic and those who visit. During his term, Harper visited Arctic communities regularly and frequently proclaimed both the importance of the North for Canada and his personal commitment to exercising Canadian sovereignty over Arctic waters. Yet, he could dismiss the evidence of climate change gathered by means of both traditional and scientific method without considering what it meant to people living in the Arctic and relying on those same waters. Many Canadian writers, politicians, and celebrities claim to love the Arctic and assume an authoritative voice on Arctic matters based on traveling there; however, the former Prime Minister's example demonstrates how inadequate authority on Arctic matters based on visiting the region can be. Without the knowledge offered by those who live on the land, the temporary visitor or tourist lacks the understanding that can be gained by studying the evidence, including the testimony of the Indigenous people of the Arctic.

These issues are apparent in James Raffan's *Circling the Midnight Sun* which emphasizes the problems in modern northern societies in order to make the point that greater autonomy is needed for people living in the Arctic to survive. A staunch ally of those living in the North, Raffan is a frequent visitor to the region, and the stated goal of his travel narrative is to give voice, or at least make space for the voices of northern people, a goal he returns to when explaining specific aspects of Arctic geography, history, and politics. Throughout, Raffan narrates his discussions with the people he meets, many of them Indigenous people, acting as a translator or mediator between cultures. This technique allows him to shape the narrative to fit his themes as well as to have the last word in many of the conversations. It also allows him to compartmentalize issues such as suicide, poverty, and addiction into discreet problems requiring action. In this portrait of Arctic societies, "global warming," he concludes, "is the least of their worries" (Raffan 3).

Sheila Watt-Cloutier seems to disagree. For her, the environment, including climate, sustains all aspects of life, and everything is interconnected. Understanding this fact is essential to understanding the Arctic. Like many popular books published by commercial presses today, *Circling the Midnight Sun* does not reference its research and its authority seems to reside in the author's frequent visits to the Arctic. The book has no index, bibliography, or notes, and the lack of this basic textual scaffolding makes it a very different book from Watt-Cloutier's which embeds evidence and references in the body of the text. Unlike Watt-Cloutier, who views climate change as a process that is affecting every aspect of Inuit life, Raffan's way of reading Arctic societies maintains the "two worlds" stereotype whereby modern life brings detrimental changes and traditional culture offers an alternative. Watt-Cloutier does not make such a separation between traditional or Indigenous Knowledge and Modernity, nor does she imply any opposition or hierarchy; instead, she addresses the aspects of tradition and Modernity coexisting in the present day because the Arctic is her homeland, and she lives it.

The Right to Be Cold describes a new methodology based on communication between Inuit living on the land and those studying the environment. This methodology not only produces better, more complete conclusions about the data, but also provides a model of governance that has the potential to decolonize. Placing Indigenous knowledge on an equal footing with scientific knowledge promotes a relationship based on equality and partnership that undermines the colonial relationship. From the experience of lobbying for global action on carbon emissions, new partnerships have formed. While there have been clashes between conservationists and hunting cultures, there is more evidence of beneficial cooperation than conflict. These conflicts, however, offer support to Watt-Cloutier's insistence on the importance of education. *The Right to Be Cold* shows readers an Inuit way, educating as a means of persuading. The value of persuasion can be observed in the rhetorical structure crafted throughout the book. The prose maintains a balanced style throughout. This is most clear when serious criticism needs to be expressed. In such cases, Watt-Cloutier uses a three-part structure: a positive statement precedes the critical statement and another positive statement follows it. Her critical comments are never harshly expressed or stark, couched as they are in constructive terms. The effect is a voice that is thoughtful and firm even when confronting opposition. This prose style articulates the style of leadership Watt-Cloutier values and practices: one based on listening to and working with others.

In *Ice and Water: Politics, Peoples, and the Arctic Council,* John English describes Watt-Cloutier's "brilliant" testimony before the U.S. Senate Committee on Commerce, Science, and Transportation chaired by Senator John McCain as "a triumph of the movement towards circumpolar collaboration" (280). It certainly was that. The knowledge needed to make good, informed decisions about the environment relies on collaboration between Indigenous and scientific ways of knowing. With their circumpolar and global outlook, the voices of Indigenous peoples are changing how the Arctic is written. Inuit are working to decolonize their homeland using Indigenous knowledge and technology, even as the land and the ice shifts beneath their feet. While representatives of the federal government continue to emphasize the consequences of global warming for security and sustainable development and to view the issue through the lens of Canadian sovereignty, Inuit representatives are dedicated to preserving the Arctic environment as the basis of their sovereignty as Indigenous people. In "Living Inuit Governance in Nunavut," Jackie

Price explains that the challenges facing Inuit people are unprecedented: "[c]limate change and resource development are changing the land physically, and this is changing the relationship Inuit have with the land" (Price 127). Resource extraction and climate change have opened a "gap between Inuit physical and conceptual experiences" that has created a "new chaos" exacerbating the challenges outlined by Ipellie's essay (Price 128–129). As the image of a permanently frozen land is giving way to the toxic, melting, or vanishing Arctic, there is a call for Indigenous knowledge even as climate change is forcing northern peoples to reinterpret what they know. The most pressing challenge facing northern communities is the impact of technological and environmental change on the northern environment, and as Watt-Cloutier's negotiating demonstrates, environmental issues such as climate change are global problems that call for working across national boundaries across the circumpolar world. Understanding these bodies of knowledge as complimentary and in dialogue, Watt-Cloutier illustrates the uniqueness of Inuit culture and within it, the importance of Indigenous Knowledge as the means by which it is conveyed from generation to generation. The knowledge of Inuit culture and the values derived from it remain vital: "It's not history. It's a continuing contemporary way of life. And it is perfectly compatible with the modern world" (Watt-Cloutier 321).

NOTES

1. The Truth and Reconciliation Commission traveled across Canada collecting the testimony of survivors of Canada's residential schools. The final report of the commission submitted in 2014 can be accessed at www.trc.com.
2. As Hartmut Lutz observes building on McGrath's work, "the earliest forms of Inuit literature are the results of literacy programs by missionaries" (72). In collaboration with Alootook Ipellie a team of scholars including Hartmut Lutz, Renate Jütting, Ruth Bradley-St. Cyr, and Hans-Ludwig Blohm published *The Diary of Abraham Ulrikah*. Abraham's diary, which tells his mentor, the Moravian brother Elsner, of the hardship and abuses the group suffers, is filled with testimony to Christ. Abraham, whose letters to his Christian mentor form the basis of the earliest known Inuit autobiography, was one of a group of Labrador Inuit taken from Hebron to be displayed in the Hamburg Zoo. After touring several European cities, the small group gradually succumbed to smallpox.
3. As the editors of the new edition explain, even the spelling of Freeman's name "Mini" was altered to the anglicized "Minnie" in the edition available when Blake was writing.

4. *Sanaaq* has been hailed as the "first Inuit novel," a distinction once given to *I, Nuligak* in order to distinguish it from oral tradition collected in books by Agnes Nanogak, Zebedee Nungak, William Oquilluk and others. For a contemporary oral history, see *Uqalurait* edited by John Bennett and Susan Rowley.

5. Melanie Magrath tells the story of the RCMP's destruction of the sled dogs in *The Long Exile*, and Sheila Watt-Cloutier references the documentary films *Qimmit: A Clash of Two Truths* (NFB, 2010) and *Echo of the Last Howl* (Taqramiut, 2004) as well as the eyewitness accounts given in the Qikiqtani Truth Commission Reports.

REFERENCES

Bennett, John, and Susan Rowley, eds. 2004. *Uqalurait: An Oral History of Nunavut.* Kingston: McGill-Queen's University Press.

Blake, Dale S. 2000. *Inuit Autobiography: Challenging the Stereotypes.* Dissertation, University of Alberta.

Bravo, Michael. 2000. Cultural Geographies in Practice: The Rhetoric of Scientific Practice in Nunavut. *Ecumene* 7 (4): 468–474.

———. 2009. Voices from the Sea Ice. *Journal of Historical Geography* 35 (2): 256–278.

Cameron, Emilie. 2015. *Far Off Metal River: Inuit Lands, Settler Stories, and the Making of the Contemporary Arctic.* Vancouver: University of British Columbia Press.

Canada Reads, 28 Mar. 2017. Accessed March 28, 2017. www.cbc.ca/books/canadareads

Echo of the Last Howl/L'écho du dernier cri. 2004. Dir. Claude Grenier and Guy Fradette. Makivik Corporation/Taqramiut Productions.

English, John. 2013. *Ice and Water: Politics, Peoples, and the Arctic Council.* Toronto: Penguin.

Fienup-Riordan, Ann. 2003 [1990]. *Eskimo Essays: Yup'ik Lives and How We See Them.* New Brunswick, NJ: Rutgers University Press.

Freeman, Mini Aodla. 2015. In *Life Among the Qalunaat,* ed. Keavy Martin, Julie Rak, and Norma Dunning. Winnipeg, MB: University of Manitoba Press.

Ghosh, Amitav. 2016. *The Great Derangement: Climate Change and the Unthinkable.* Chicago: University of Chicago Press.

Haraway, Donna J. 2016. *Staying with the Trouble Making Kin in the Chthulucene.* Durham, NC: Duke University Press.

Harper, Kenn. 2007. Inuk Felt Tortures of Transition, then Relived them Through the Arts. *Globe and Mail,* 28 November, S9.

Henderson, Heather. 1988. North and South: Autobiography and the Problems of Translation. In *Reflections: Autobiography and Canadian Literature,* ed. K.P. Stich, 61–68. Ottawa: University of Ottawa Press.

Ipellie, Alootook. 2001. People of the Good Land. In *The Voice of the Natives: The Canadian North and Alaska*, ed. Hans-Ludwig Blohm, 19–31. Ottawa: Penumbra.

Leduc, Timothy B. 2010. *Climate, Culture, Change: Inuit and Western Dialogues with a Warming North*. Ottawa: University of Ottawa Press.

Magrath, Melanie. 2007. *The Long Exile: A Tale of Betrayal and Survival in the High Arctic*. New York: Vintage.

Martin, Keavy, and Taqralik Partridge. 2016. 'What Will Inuit Think?' Keavy Martin and Taqralik Partridge Talk Inuit Literature. In *The Oxford Handbook of Canadian Literature*, ed. Cynthia Sugars, 191–208. New York: Oxford University Press.

McCannon, John. 2012. *A History of the Arctic: Nature, Exploration and Exploitation*. London: Reaktion.

McGhee, Robert. 2004. *The Last Imaginary Place: A Human History of the Arctic World*. Toronto: Key Porter.

McGrath, Robin. 1984. *Canadian Inuit Literature: The Development of a Tradition*. Ottawa: National Museums of Canada.

———. 1993. Samuel Hearne and the Inuit Oral Tradition. *Studies in Canadian Literature* 18 (2): 94–109.

Nanogak, Agnes. 1972. *Tales from the Igloo*. Edmonton: Hurtig.

———. 1986. *More Tales from the Igloo*. Edmonton: Hurtig.

Nappaaluk, Mitiarjuk. 2014 [1983]. *Sanaaq: An Inuit Novel*. Translated from Inuktitut by Bernard Saladin D'Anglure. Translated from French by Peter Frost. Winnipeg, MB: University of Manitoba Press.

Nuliajuk: Mother of the Sea Beasts. 2001. Dir. John Houston and Peter d'Entremont. Triad Films.

Nuligak. 1966. *I, Nuligak*. Translated by Maurice Metayer. Toronto: Peter Martin.

Nungak, Zebedee, and Eugene Arima, eds. 1969. *Unikkaatuat sanaugarngnik atyingnaliit Puvirngni turngmit/Eskimo Stories from Povungniyuk, Quebec*. Ottawa: Queen's Printer.

Oquilluk, William A., and Laurel L. Bland. 1981. *People of Kauwerak: Legends of the Northern Eskimo*. Anchorage: Alaska Pacific University Press.

Price, Jackie. 2008. Living Inuit Governance in Nunavut. In *Lighting the Eighth Fire: The Liberation, Resurgence, and Protection of Indigenous Nations*, ed. Leanne Simpson, 127–138. Winnipeg, MB: ARP.

Qapirangajuq: Inuit Knowledge and Climate Change. 2009. Dir. Zacharias Kunuk and Ian Mauro. Isuma Productions. Accessed May 5, 2017. www.isuma.tv/inuit-knowledge-and-climate-change/movie

Qimmit: A Clash of Two Truths. 2010. Dir. Ole Gjerstad and Joelie Sanguya. National Film Board of Canada. Accessed December 5, 2016. www.nfb.ca/film/qimmit-clash_of_two_truths/

Raffan, James. 2014. *Circling the Midnight Sun: Culture and Change in the Invisible Arctic*. Toronto: HarperCollins.

Sangster, Joan. 2016. *The Iconic North: Cultural Constructions of Aboriginal Life in Postwar Canada*. Vancouver: University of British Columbia Press.

Sejersen, Frank. 2015. *Rethinking Greenland and the Arctic in the Era of Climate Change*. London: Routledge.

Ulrikah, Abraham. 2005. *The Diary of Abraham Ulrikah*. Edited and translated by Hartmut Lutz, Renate Jütting, Ruth Bradley-St. Cyr, and Hans-Ludwig Blohm. Ottawa: University of Ottawa Press.

Vance, Jonathan F. 2002. *High Flight: Aviation and the Canadian Imagination*. Toronto: Penguin.

Watt-Cloutier, Sheila. 2015. *The Right to Be Cold: One Woman's Story of Protecting her Culture, the Arctic, and the Whole Planet*. Toronto: Allen Lane.

Epilogue

Writing in the Anthropocene, the artist must search for new forms or revive ancient ones that can convey the varied yet shared experience of climate change. By addressing the future of the Arctic, the two writers discussed in this book are also working against the "great derangement" described so compellingly by Amitav Ghosh. Coming from very different backgrounds and lived experience as an Inuit leader and environmental activist and a Newfoundland writer who as a child emigrated from Britain respectively, Sheila Watt-Cloutier and Kathleen Winter embrace the non-fictional memoir, a form that focuses on a significant period or set of events in the life of the writer but that can also accommodate as broad a frame of reference as needed. The ancient knowledge of Indigenous peoples, the data and evidence of environmental science, and the political landscape of international relations can all be held within the frame. The result is literary knowledge of the environmental crisis caused by global warming.

This book has been about imagining the Arctic in the changing world of the Anthropocene, a world 'after nordicity' and beyond grand narratives of Arctic discovery represented by the Franklin expedition. As the grand narratives of Arctic discovery are being replaced by gradual discovery of Indigenous knowledge, a new form of Arctic discourse based on active listening and respect seems to be emerging, though it is still too early to predict its impact. In the Canadian context, these books seem to indicate a movement toward recognition of Inuit sovereignty and away from preoccupations with Canadian nationalism in the Arctic. As Indigenous

© The Author(s) 2018
R. Hulan, *Climate Change and Writing the Canadian Arctic*,
https://doi.org/10.1007/978-3-319-69329-3

people assert their sovereignty in economic, political, and social terms, Sheila Watt-Cloutier's work is navigating what Pauline Wakeham calls "the vital interplay between the ostensibly 'soft' domain of cultural politics and the 'hard sphere' of state geopolitics and sovereignty-claiming" (88). Having worked at the highest levels of international diplomacy, Watt-Cloutier has turned to writing as a means to further her work within the cultural domain, teaching readers as she does. In her memoir of Arctic travel, Kathleen Winter shows herself to be a willing student with a growing awareness of her privilege as a settler even though by the end of the book, her education has barely begun. *The Right to Be Cold* offers a set of teachings that articulates, explains, and enacts Inuit sovereignty by demonstrating how all aspects of lived experience are connected. For the reader who wants to learn, it teaches what Inuit know and value, and how knowledge is shared and conveyed. One of the lessons learned from carefully and actively reading Indigenous writers is that Modernity, like climate change, is a shared condition. The image of the Arctic as a margin awaiting the arrival of Modernity has been shattered by the knowledge that the colonization of the North, and access to its resources of both goods and labor, its strategic importance, and symbolic value, have contributed to making Modernity what it is and to the unequal power relations that continue to shape who speaks and who is heard. And just as Modernity is shared, so too are its consequences of which climate change is perhaps the most dire. In the right form, literature has a part to play in facing these consequences by addressing the environmental crisis caused by global warming. Both writers show how the future of the Arctic and the earth as a whole depends on greater understanding of Indigenous knowledge, including literary knowledge, of the Arctic world.

REFERENCES

Ghosh, Amitav. 2016. *The Great Derangement: Climate Change and the Unthinkable*. Chicago: University of Chicago Press.

Wakeham, Pauline. 2014. At the Intersection of Apology and Sovereignty: The Arctic Exile Monument Project. *Cultural Critique* 87: 84–143.

Watt-Cloutier, Sheila. 2015. *The Right to Be Cold: One Woman's Story of Protecting her Culture, the Arctic, and the Whole Planet*. Toronto: Allen Lane.

Winter, Kathleen. 2014. *Boundless: Tracing Land and Dream in a New Northwest Passage*. Toronto: Anansi.

INDEX[1]

[1] Note: Page number followed by "n" refers to notes.

© The Author(s) 2018
R. Hulan, *Climate Change and Writing the Canadian Arctic*,
https://doi.org/10.1007/978-3-319-69329-3

CPSIA information can be obtained
at www.ICGtesting.com
Printed in the USA
BVHW01*1909220118
505985BV00006B/46/P